AROMATIC FLUORINATION

NEW DIRECTIONS in ORGANIC and BIOLOGICAL CHEMISTRY

Series Editor: C.W. Rees, CBE, FRS
Imperial College of Science, Technology and Medicine, London, UK

Published and Forthcoming Titles

AROMATIC FLUORINATION

James H. Clark
David Wails
Tony W. Bastock

CRC Press
Boca Raton   New York   London   Tokyo

Developmental Editor:	Felicia Shapiro
Assistant Managing Editor:	Paul Gottehrer
Marketing Manager:	Arlene Massey
Direct Marketing Manager:	Becky McEldowney
Cover design:	Denise Craig
PrePress:	Kevin Luong
Manufacturing:	Sheri Schwartz

Library of Congress Cataloging-in-Publication Data

Catalog record is available from the Library of Congress.

Preface

Aromatic fluorine chemistry has a history that dates back over 100 years. Industrial processes for the manufacture of fluoromatics have been in operation for over 50 years but it is only in more recent years that the commercial potential for aromatic fluorine compounds has begun to be realized. Three significant developments relevant to the growth of commercial interest in aromatic fluorine compounds are (1) the first efficient "halex" methods for the preparation of selectively fluorinated aromatics which were reported in the 1960s; (2) the appreciation of the significant changes in the properties of aromatic molecules that can result from only low levels of fluorine incorporation; and (3) the increasing commercial importance of fine and specialty chemicals as markets for pharmaceuticals, agrochemicals, electronic chemicals and other high value chemical products have grown. These developments have resulted in the emergence of a rapidly increasing number of selectively fluorinated aromatic molecules based on a growing number of small fluorine-containing aromatic substituents and the associated need for new and more efficient methods of synthesis.

Some measures of the commercial value of aromatic fluorine compounds are the world consumption of ca. 4.5×10^3 tonnes per annum and the remarkable number of patents associated with their manufacture or use (ca. 13,000 between 1981 and 1986). While the total world market may seem small in comparison to that for the more established highly fluorinated aliphatics, the figure is deceptive and the growth potential is considerable. The market for fluorine-containing pharmaceutical *products* is over $10 billion based on some 500 products with one drug alone having world sales of ca. $1.2 billion. The market for one agrochemical *product* alone was worth $400 million at its peak and the growth in sales for fluorine-containing agrochemicals is now expected to be ca. 7% per annum. Fluorine-containing reactive dyes are believed to hold about 20% of the world market and at the very low volume but high added value end, fluorine is now the element par excellence for achieving useful effects in liquid crystals. new applications for aromatic fluorine compounds include high performance polymers and advanced materials. As new products involving alternative fluorine-containing substituents and multiple substituted aromatics become available and as the markets for higher value products and intermediates expand, aromatic fluorine chemistry will continue to grow and develop. The historical landmarks in aromatic fluorine chemistry, the effects that small amounts of fluorine can have on the properties of aromatic molecules, the applications for the compounds, and their synthesis and manufacture are reviewed in Chapter 1.

In the last 30 years, halex chemistry involving the direct replacement of aromatic chlorine by fluorine has grown into the most important method for the preparation of selectively fluorinated aromatic molecules. This superficially simple reaction is complicated by the poor reactivity of traditional sources of nucleophilic fluorine and this has led to an enormous research effort to enhance the reactivity of simple metal fluorides, to design catalytic systems, and to prepare more reactive fluorides. The subject is also developing through the extension of nucleophilic fluorination to fluorodenitration as an alternative commercial method and through the search for more environmentally benign methods of manufacture such as no-solvent processes. The halex and related methods for preparing selectively fluorinated aromatic molecules are described in Chapter 2.

The Baltz-Schiemann and other diazonium-based methods of fluoroaromatic synthesis are well established in fluorine chemistry and the original methods developed in the early part of the century are still valid today. Diazonium reactions are generally multi-stage and require careful control but the first industrial process for the continuous manufacture of a fluoroaromatic is based on such chemistry. The methods of fluoroaromatic synthesis via diazoniums, the mechanism of the reactions, and the scope of the chemistry are discussed in Chapter 3.

Fluorine chemistry is as noteworthy for the remarkable range of inorganic fluorinating agents as for the breadth of organofluorine products. Reactive reagents for selective fluorination such as N-F and O-F compounds and electrochemical synthesis of low-fluorinated compounds offer much promise as alternatives to the more difficult to control aromatic fluorinations based on more traditional methods such as elemental fluorine and high valency metal fluorides. These and other alternatives to halex and diazonium routes to fluoroaromatics are described in Chapter 4.

The trifluoromethyl group is second only in importance to fluorine itself as a substituent in aromatic fluorine chemistry. Its value is already established in pharmaceutical and agrochemical products and its importance in other areas such as high performance polymers is beginning to emerge. Traditional methods of manufacture based on acid-catalyzed halogen exchange remain viable but new methods including those based on organometallic reagents are likely to play an increasingly important role in the future. The properties of trifluoromethylated aromatics and the most important methods of preparing the compounds are reviewed in Chapter 5.

The very high lipophilicity of the trifluoromethylthio group makes it a particularly valuable substituent in the design of pharmacologically active compounds and some highly active aromatic pharmaceutical and agrochemical products containing this group are known. The trifluoromethylthio group and the trifluoromethylsulfonyl group in particular have strong electronic effects on the properties of molecules and these, along with the good physical and chemical stabilities of the groups, can be usefully exploited in the design of effective chemicals. Routes to trifluoromethylthioaromatics and trifluoromethylsulfonylaromatics can involve radical or electrophilic chemistry and to some extent nucleophilic substitution and the choice of the most appropriate method is often far from clear. The various methods of preparation along with the key properties of the product molecules are discussed in Chapter 6.

As the search for new products continues so will the need for designing new selectively fluorinated aromatic molecules. The availability of viable routes to aromatic substituents other than F, CF_3, and CF_3S will become increasingly important as will an understanding of the properties of the resulting product molecules. The most important routes to aromatic molecules containing some of the most promising alternative substituents such as OCF_3 and perfluoroalkyl groups larger than CF_3 are described in Chapter 7.

At the time of this writing there was an over capacity in aromatic fluorine compounds resulting in part from a global recession but also from a rationalization of the industry which has seen several companies leave the market. Changes in technology resulting from a trend towards continuous processes and the need to reduce the environmental impact of some manufacturing processes will also have an increasing effect on the industry. The projected growth in demand for agrochemical products, the expanding markets for fine and specialty chemicals and the need to reduce doseage (hence requiring higher molecular activity) in many applications are all likely to result in real growth for aromatic fluorine compounds in the future. The supply and demand for these compounds, the structural types of products on the market and in development, the manufacturing methods and the synthetic strategies for aromatic fluorine compounds are described in Chapter 8.

This is not meant to be a comprehensive guide to aromatic fluorine chemistry. It is, however, meant to assist both the newcomer and the specialist in identifying the best methods for preparing aromatic fluorine compounds where the fluorine is separated by no more than two atoms from the ring. It should also help the nonspecialist and indeed the uninitiated to recognize and appreciate the value of fluorine in terms of its useful effects on the properties of aromatic molecules. The range of commercial examples should serve to demonstrate that what is often regarded as a rather unpredictable, even dangerous element can be tamed to create the most sophisticated effects and to produce products ranging from antibiotics to liquid crystal displays that are for the benefit of humanity.

James Clark
York, March, 1996

The Authors

James H. Clark is currently Professor of Industrial and Applied Chemistry at the University of York, England. He obtained his B.Sc. and Ph.D. degrees at Kings College, London. Professor Clark has an international reputation for his research on clean synthesis and fluorine chemistry and he has been involved in national and international award winning projects. He also a Royal Society of Chemistry medal winner. Professor Clark is the author of over 100 original research articles and two previous books.

David Wails is a Research Chemist currently based at Brock University in Canada where he is working on new catalytic methods of synthesis. He obtained his B.Sc. at the University of York, England where he also carried out research on fluorinated aromatic polymers that led to his Ph.D.

Tony W. Bastock is currently Group Sales and Marketing Director of Contract Chemicals Ltd. and Director in Charge of Contract Catalysts. He was educated at the University of Birmingham where he obtained a B.Sc. in chemistry and a Ph.D. in fluorine chemistry. He is the chairman of the University-Industry collaborative research project on new environmentally-friendly catalysts, "Envirocats", that has won the prestigious Clean Technology awards from the Royal Society of Arts and the European Community, and has been featured on television and radio. He was the winner of the 1994 Royal Society of Arts Howarth Medal for Enterprise and Innovation in the North West.

Table of Contents

Chapter 7
Other Aromatic Ring Substituents 139

Chapter 8
Industrial Aspects of Aromatic Fluorine Chemistry 159

Chapter 1

Introduction to Aromatic Fluorination

1.1 HISTORICAL DEVELOPMENTS

Aromatic fluorine chemistry has a remarkably long history and dates back to the first successful syntheses of aryl C-F bonds in 1870.[1] These first syntheses were a direct consequence of the earlier discovery of aromatic diazo compounds which opened the door to many aryl derivatives, including the fluorides:

$$ArH \rightarrow ArNO_2 \rightarrow ArNH_2 \rightarrow ArN_2^+ \rightarrow ArF$$

Routes to fluoroaromatics based on the preparation and subsequent decomposition of diazonium intermediates have played an important part in the subject ever since.

The synthesis of fluorobenzene itself took a little longer. Indeed the first claim to the preparation of the compound, from calcium fluorobenzoate, proved to be erroneous and actually gave phenol,[2] thus representing the first example of nucleophilic displacement of fluorine! Fluorobenzene was first synthesized by desulfonylation of potassium p-fluorobenzene sulfonate.[2]

Phenols also complicated the original diazonium-based routes to fluoroaromatics. The use of aqueous rather than anhydrous hydrogen fluoride led to reduced fluoride nucleophilicity and competitive hydroxide nucleophilic chemistry. The first method for countering this problem involved the use of diazopiperidides and this led to the preparation of numerous fluoroaromatics, including 1,4-difluorobenzene.[3] Their availability led in turn to the first electrophilic substitutions on fluorobenzene and the successful development of methods of analysis for fluoroaromatics.

Significant developments in the area in the early part of the 20th century[3] included the appreciation that the aromatic fluorine could be susceptible to nucleophilic substitution (and hence the synthesis and application of Sangers' reagent, for example), the extension of the single-ring systems to the synthesis of fluorobiphenyls, fluoronaphthalenes, and fluoroanthraquinones, and most notably the discovery of the Balz-Schiemann reaction.

The Balz-Schiemann reaction is based on the original discovery that the earlier reported arene diazonium tetrafluoroborates,[4] when heated, gave good yields of the corresponding fluoroaromatics:[5]

$$ArNH_2 \rightarrow ArN_2^+X^- \rightarrow ArN_2^+BF_4^- \rightarrow ArF + N_2 + BF_3$$

This proved to be a general method for the preparation of very many fluoroaromatics and, as importantly, it enabled the synthesis of fluoroaromatics in any well-equipped laboratory,[6-9] so that organofluorine chemistry became familiar to many aromatic organic chemists (whereas aliphatic organofluorine chemistry remained a highly specialized area). Interestingly it was only after the Balz-Schiemann reaction became established that it was reported that the original method of fluoroaromatic synthesis based on the diazonium fluoride could be made efficient by the use of anhydrous hydrogen fluoride. The two methods remain alternatives, with the tetrafluoroborate route having the advantage of involving a relatively stable salt intermediate, whereas the fluoride route is less expensive.

The synthesis of 2,4-dinitrofluorobenzene by halogen exchange using potassium fluoride was another major breakthrough in the area and represented the beginning of what was to become a major industrial process and a great rival to the diazonium-based methods for the synthesis of selectively fluorinated aromatic molecules.[10]

The large majority of fluoroaromatics reported in the first 40 years of research on the subject were benzene derivatives, although a few polycyclics (including biphenyls, anthraquinones, and naphthalenes) were reported. The first fluorinated heterocycle can be traced back to 1915 with the reported synthesis of 2-fluoropyridine.[11] Aromatic fluorination was limited to one or two fluorines in a ring and attempts to extend this using diazonium chemistry revealed the increasing difficulty of progressive fluorination using this methodology. More extensive fluorination had to wait for the development of exhaustive methods in the 1940s.

Perfluorobenzene was first reported in 1947,[12] although it had actually been synthesized in the previous decade.[13] This was quickly followed by the first perfluorinated heterocycle, pentafluoropyridine, and subsequently a large number of derivatized perfluoroaromatics which enabled the establishment of directional effects in nucleophilic substitution reactions of these molecules. The commercialization of polyfluoroaromatic chemistry relied on cobalt trifluoride as the "tamed" source of fluorine; but, this was replaced by halogen exchange technology based on the use of the perchloro-substrates.[14]

The exchange of chlorine by fluorine also opened the door to the synthesis of trifluoromethylaromatics, although this process, unlike the reactions of chloroaromatics with fluorides, is acid catalyzed. This so-called Swarts process continues to form the basis of the synthesis of the very many commercial benzotrifluoride derivatives.[11] The same technology can also be used to synthesize other important small fluorine-containing substituents, notably SCF_3 and OCF_3. These substituents are rapidly gaining in popularity either because of increased effect or to enable "patent jumping". New, mostly organometallic-based synthetic methods into C-1 and other small fluorine-containing substituents offer much hope for the future although industry will be slow to respond.

The subject, which has its origins in the pioneering and highly problematic research of the second half of the 19th century, is now well established as a subject and an industry in its own right. The volume of scientific and technical literature on organofluorine chemistry is staggering and the growth of academic interest and industrial exploitation is extraordinary. These statements can be exemplified in many ways:

- Of the 10 million compounds registered in the American Chemical Society's (ACS) Chemical Abstracts in 1990, 6.2% are compounds containing C-F bonds; the number of documented novel organofluorine compounds has increased from about 12,000 in 1970 to a remarkable 60,000 in 1990; many of these compounds are reported in patents.
- There has been a steady increase in the number of papers on organofluorine chemistry published each year — from about 3000 in 1975 (approximately 1% of all papers published in that year) to about 4000 in 1984 (1.2%), to about 5000 in 1990 (1.3%).
- The large proportion of conference papers concerned with fluorine (typically about 10% at recent ACS meetings) has led to a sudden increase in the number of international symposia on aspects of fluorine and organofluorine chemistry in particular. The regular International and European Fluorine Symposia are now supplemented by numerous specialist meetings, such as those devoted to fluorine in the context of polymers (e.g., ACS, 1990), mass spectrometry (ACS, 1990), coatings (UK, 1994), medicine (UK, 1994), and agriculture (UK, 1995).
- Fluorine is now a major element in the research and development programs of several major industrial sectors, including pharmaceuticals, agricultural chemicals, polymers, liquid crystals, surface coatings, solvents, propellants, fluids, dyes, blood substitutes, anesthetics, and others.
- It has been suggested that a conservative estimate of the value of commercial products containing fluorine is $50 billion per year.

While aromatic fluorine compounds represent only a part of this, that part is undoubtedly very significant and likely to grow. Many of the traditional outlets for organofluorine compounds (propellants, refrigerants, coolants, solvents, etc.) are now being challenged as a result of environmental considerations and the volume of business may well decrease, at least in the medium term. Aromatic fluorine compounds have largely avoided the bad publicity that (quite incorrectly) all volatile halogen-containing compounds have been tarred by. Selectively fluorinated aromatic compounds are rapidly gaining popularity as low dosage drugs and crop protection agents, among other applications. The increasing need for novelty and for enhanced or special effect will guarantee the continued search for new compounds and with it a thriving literature and academic and industrial research community.

1.2 PROPERTIES OF FLUORINATED AROMATIC MOLECULES

It has been known for many years that fluorine can have profound effects on molecular activity. It is also now recognized that high levels of fluorination are not necessary for useful activity effects to be achieved. Selective fluorination is an extremely effective tool for modifying and probing molecular reactivity. The unique characteristics of the C-F bond have resulted in highly significant advances in many areas including electronics, medicine, agriculture, health care, and materials, and selectively fluorinated aromatics have an important current and future role in all of these areas.

1.2.1 Fundamental Properties

The typical characteristics of organofluorine compounds are given in Table 1.1. The "fluorine factor" in organofluorine chemistry arises from a unique combination of properties associated with the atom and its bond to carbon, its high electronegativity, the remarkably strong C-F bond, and the relatively small volume occupied by the most interesting and valuable organofluorine substituents, F, CF_3, CF_3S, and CF_3O, among others. Size factors can be misleading, however, and the often quoted similarity between F and H is deceptive. The volume occupied by a CF_3 group is much larger than that for a CH_3 group and in fact is more similar to a $(CH_3)_2CH$ group. In fact in areas such as drug design where new aryl substituents need to be very similar, in terms of steric demands, to the groups they are replacing, it is often better to regard F and OH as chemical isosteres. This is nicely illustrated by the widespread ability of C-F to mimic C-OH groups in bioactive molecules (Table 1.2).

Table 1.1 Typical Characteristics of Organofluorine Compounds

Properties	Characteristics
Electronegativity	Fluorine is the most electronegative element; other small fluorine-containing groups (CF_3, CF_3O, CF_3S, etc.) are also strongly electron withdrawing.
Bond strength	The mean bond dissociation energy for C-H (485 kJ mol^{-1}) is significantly greater than those for C-H (413 kJ mol^{-1}) and C-Cl (339 kJ mol^{-1}). This can result in enhanced chemical and thermal stability.
Size	Fluorine has a small van der Waals radius (0.147 nm) compared to chlorine (0.177 nm) and not very different from that of hydrogen (0.120 nm). F can replace H with minimal steric disruption.
Volume	The volume occupied by an aryl fluorine (5.80 cm^3 mol^{-1}) is less than that for a hydroxyl group in a phenol (ca. 8 cm^3 mol^{-1}). The volume of a CF_3 group (21.3 cm^3 mol^{-1}) is significantly larger than that of a CH_3 group (13.7 cm^3 mol^{-1}) and a carbonyl group (11.7 cm^3 mol^{-1}) and about the same size as a $(CF_3)_2CH$ group. $(CH_3)_2CF$ is comparable in size to $(CH_3)_3C$.

Table 1.2 Van der Waals Volumes

Group	Volume/cm^3 mol^{-1}
C(1°)-F	5 · 72
C(2°, 3°)-F	6 · 20
C(aromatic)-F	5 · 80
$-CF_2-$	15 · 3
$-CH_2F$	16 · 0
$-CF_3$	21 · 3
$-O-$ (dialkyl ether)	3 · 7
$-O-$ (e.g., tetrahydrofuran)	5 · 2
$\diagdown C=O$	11 · 7
$-CH_3$	13 · 7
$-CH_2-$	10 · 3
$-OH$ (non H-bonded)	8 · 0

1.2.2 Effects of Fluorine on Chemical Reactivity

Electronic effects associated with fluorine are both obvious and unexpected. The powerful electron withdrawing ability of the most electronegative element (–I) goes hand-in-hand with the apparent electron pushing effect (+M, +I_π, see Figure 1.1), the latter being especially important in aromatic fluorine chemistry. The result of these opposing forces is that the electronic effects on molecular properties of substituting H by F are often very different to those that might have been expected.

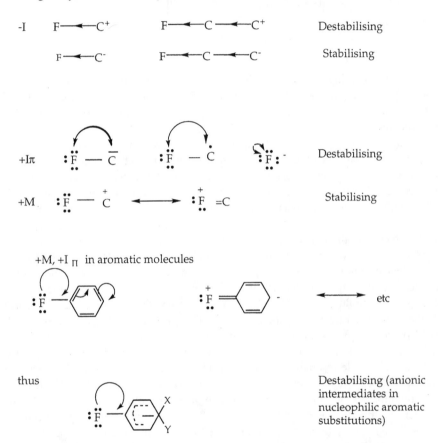

FIGURE 1.1   Electronic effects associated with fluorine in organic molecules.

Thus, in spite of its powerful –I effect, F can be a net activator towards electrophilic substitution. Fluorobenzene, for example, undergoes ortho-para electrophilic nitration (benzotrifluoride undergoes meta attack, rather more predictably). In fact fluorobenzene is the most reactive of the monohalobenzenes towards electrophilic attack:

$$C_6H_5F > C_6H_5Cl > C_6H_5Br = C_6H_5I$$

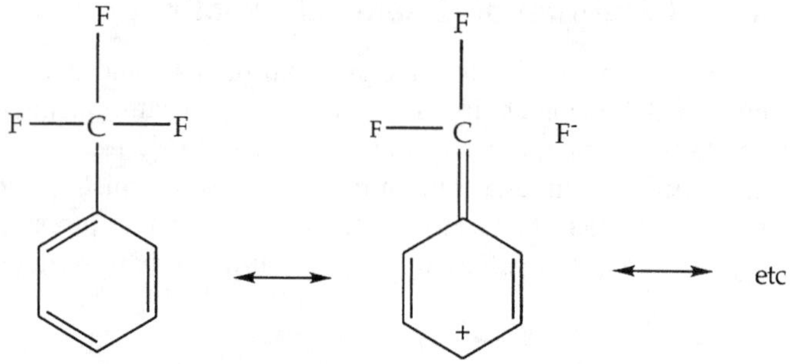

FIGURE 1.2   Fluorine hyperconjugation.

The deactivating effect of the CF_3 group towards electrophilic attack is accompanied by an activating effect of the group towards nucleophilic substitution. These rather predictable observations hide a rather more complex picture of perfluoroalkyl substituent effects.

Data derived on the basis of separating the overall effect of substituents into inductive and resonance contributions[15] indicate an apparent resonance contribution to the effect of perfluoroalkyl groups. The apparent resonance contribution of a perfluoroalkyl group is typically somewhat less than half the apparent inductive effect but nonetheless significant. It is tempting to draw an analogy with hydrocarbon systems and introduce a concept of "fluorine hyperconjugation"[16] (Figure 1.2). Several "effects" in aliphatic and aromatic fluorine chemistry have been explained on the basis of this concept, including the anomalously high dipole moments of *para*-aminobenzotrifluorides.[17] However, there are also persuasive arguments against the existence of this effect, most notably the apparent lack of dependence on the "effect" of replacing the fluorines in CF_3 by other CF_3 groups; thus, the relative importance of the resonance effect is apparently very similar in CF_3, CF_2CF_3, and $CF(CF_3)_2$. It is also increasingly difficult to envisage "fluorine hyperconjugation" with other important small fluorine-containing groups such as SCF_3 and OCF_3; yet, these groups have a measurable resonance contribution to their substituent effects as measured by their effects on the ionization constants of the appropriate benzoic acids and anilinium ions (Table 1.3).[15,18] The importance of the apparent resonance effect on the molecular properties of aromatics possessing small fluorine-containing substituents cannot be ignored, although its origins may be less than fully understood. It is apparent that such groups are capable of polarizing the pi-electrons of the aromatic nucleus and that the inductive effect is best thought of as a combination of effects on the sigma-bond framework and on the pi-electrons (reminiscent of the situation in pyridine).[16,19,20]

The effects of fluorine and fluorine-containing groups on chemical reactivity are considered in more detail in individual chapters.

1.2.3 Effects of Fluorine on Physical Properties

Fluorine substituents can have a profound effect on the physical properties of the molecule and indeed such effects are often the sole reason for the application value

Table 1.3 Substituent Parameters for Small Fluorinated Groups

	Hammett		Taft	
	σm	σp	σi	$\sigma r/\sigma_R{}^\circ$
CF_3	0.45	0.57	0.38	0.38
CF_3CF_2	0.51	0.69	0.41[a]	0.11[a]
$(CF_3)_2CF$	0.44	0.60	0.36	0.21
SCF_3	0.43	0.57	0.36	0.20
OCF_3	0.43	0.31	0.45	−0.14

Note: Data are average values from benzoic acid and anilinium ionization constants.

[a] Values from ^{19}F nuclear magnetic resonance chemical shifts.

of a fluorochemical. Effects on aromatic systems are generally rather less spectacular than in aliphatic systems where perfluorination can produce extreme effects in hydrophobicity and oleophobicity, gas solubility, and chemical stability. Generally speaking, the most important effects are those on acidity and basicity, hydrogen bonding, lipophilicity, solvent polarity, surface energies, and boiling points.[21]

The effects of fluorine and fluorine-containing substituents on the acidity and basicity of aromatic molecules (benzoic acids, phenols, anilines) is significant but not great (see, for example, Chapter 2 and in particular Table 2.3). In a few cases the effect is especially pronounced; thus, one fluorinated substituent can increase the acidity of imidazole by almost 5 pK_a units (see Figure 2.9; it is interesting to compare the similar change in acidity on going from ethanoic (acetic) acid to trifluoroethanoic (trifluoroacetic) acid).

It can also be assumed that fluorine substituents will affect the hydrogen bonding properties of aromatic molecules such as phenols, acetophenones, benzoic acids, etc. These effects are expected to parallel those on the acidities and basicities of the molecules and to have their origins in predictable electronic effects. Interestingly, despite the extremely strong hydrogen bonding properties of the fluoride ion, the aromatic C-F group is not significantly hydrogen-bonding active and very few definitive examples of its involvement in hydrogen bonding are known (see Chapter 2). Generally, covalently bonded oxygen and nitrogen will always be preferred to covalently bonded fluorine, presumably because of the high electronegativity of F, making it a poor hydrogen bond electron donor (hydrogen acceptor). It is, however, important to remember that the group (and possibly other small fluorine-containing groups) might take part in secondary associations through other interactions such as those with alkali metals (e.g., in biochemical systems). Fluorine substituents that possess hydrogens, e.g., CF_2H, CH_2CF_3, will be capable of acting as hydrogen bond proton donors themselves. These can be considered analogous to the many aliphatic organofluorine compounds that are useful inhalation anesthetics and that contain CH groups believed to be hydrogen-bonding active *in vivo*, e.g., $CF_3CHClBr$ (halothane), and $CH_3OCF_2CHCl_2$ (methoxyflurane).[22] Remarkably, hexafluorobenzene is also a potent anesthetic and is believed to function via hydrogen bonding effects. Here the anesthetic must act as a hydrogen bond proton acceptor (electron donor) via its pi-cloud,[23] although it is not clear why this should be so effective in a molecule that has six electron withdrawing substituents.

Lipophilicity is a very important parameter in fluoroaromatic chemistry. Generally, fluorine substitution enhances molecular lipophilicity and therefore increases drug efficacy.[21,23] However, the effect ranges from the very small (e.g., fluorine itself) to the very large (e.g., CF_3S) (Table 1.4). While most of the interest in this area has come from the pharmaceuticals sector, there is a growing interest in the effects of fluorine substituents on the hydrophobic/oleophobic surface properties of polyaromatics (see Chapter 5). Enhanced atmospheric resistance of polymers can be useful in areas such as electronic circuit board coatings and aerospace applications.

Table 1.4 Hydrophobic Parameters
(1-octanol-water Partition Coefficients)

OCH_2CO_2H

X	π^a
OH	−0.49
H	0
NO_2	+0.11
F	+0.13
CH_3	+0.51
Cl	+0.76
SO_2CF_3	+0.93
CF_3	+1.07
OCF_3	+1.21
SCF_3	+1.58

[a] Hydrophobic parameter, defined as $\pi x = \log P_x - \log P_H$ where P_x is an oil/water partition coefficient (octan-1-ol/water) of a derivative and P_H that of a parent compound.

Fluoroaromatic solvents are very much the poor relation to fluoroaliphatic solvents, perhaps because of the relatively small effects that at least small numbers of fluorines can have on aromatic solvent polarity and because of the more limited range of highly volatile fluoroaromatics. Benzotrifluoride and hexafluorobenzene are sometimes used as solvents, especially for other fluorinated compounds.

The effects of fluorine substituents on other molecular physical properties are often small and are described in individual chapters, e.g., Chapter 2 (see Table 2.3). Effects on electronic properties are very important in the context of liquid crystals, an industrial sector in which fluorine has had a dramatic effect (see Chapter 5).

1.3 APPLICATIONS FOR FLUORINATED AROMATICS[24-26]

The commercial exploitation of fluoroaromatics is a more recent phenomenon than fluoroaliphatics, although the first report on the biological effects of fluorine substitution on aromatics can be traced back to when fluorobenzoic acids were fed to dogs, leading to the excretion of fluorohippuric acids.[27] The first fluorine-containing drugs were fluorosteroids.[28] Fluoroaromatics have become key building blocks in the pharmaceutical, agrochemical, dyes, and advanced materials areas of so-called "effect" chemicals. These applications clearly demonstrate the unique benefits that fluorine can bring to mankind and perhaps go some way towards alleviating the bad press that the element seems to have suffered through its association with chlorine (which is ironic, when the ozone-depleting effects of chlorofluorocarbon, for which replacements are now so eagerly sought, are entirely due to the C-Cl bond).

The reasons for the rapid growth of interest in fluoroaromatic building blocks are essentially that effect chemicals containing these units show increased efficacy over the nonfluorinated analogues. This can manifest itself as lower dosages, broader spectrum of activity, lower toxicity, and unique physical properties. Examples of important fluoroaromatic building blocks are given in Table 1.5 along with the associated fluorine technologies. Halogen exchange and diazotization are very important synthetic methods for making such molecules. These techniques are discussed further in Chapters 2 and 3. Examples of products derived from these and further discussion on their applications can be found in Chapter 8.

The importance of fluorine to the agrochemical field can be judged from the remarkable fact that about one half of the patent applications in recent years in this sector involve organofluorine compounds. Of particular importance in this context are the advantages of lower dosage, increased selectivity, and crop safety that fluorine can provide. Fluoro organics find applications as fungicides, insecticides, and herbicides and the wide range of active structures makes it difficult to see any common features that could help structural prediction. Some examples of fluorine-containing pesticides are given in Figure 1.3 and it is particularly interesting to note the wide range of fluorine-containing substituents, including single-ring fluorine, multiple-ring fluorines, CF_3, OCF_3, SCF_3, and CHF_2.

Fluorine-containing pharmaceuticals have made major contributions in several important areas including anticancer drugs (e.g., 5-fluorouracil), anti-inflammatory agents (e.g., fluorosteroids), antimalarials, tranquilizers and relaxants, and anesthetics (e.g., halothane). A related application is via positron emission transaxial tomography (PETT), which is a noninvasive medical diagnostic tool considerably enhanced by the availability of the positron emitter ^{18}F. The half-life of ^{18}F is significant (110 min) and allows transport over reasonable distances when compared to alternatives (e.g., ^{11}C). PETT can be used for studying metabolic pathways in living systems and here the ability of fluorine to replace hydrogen with little change in steric demands of the molecule is very important: rapid synthesis is required here and fluorodenitration methods are proving popular (see Chapter 2). On a smaller scale, the availability of fluoride as an anionic leaving group enables the facile use of 1-fluoro-2,4-dinitrobenzene (Sangers' reagent) as a reagent for labeling peptides and

Table 1.5 Some Important Fluoroaromatic Building Blocks

Building block	Fluoroaromatic technology commonly associated with manufacture
Br—⟨⟩—F	Balz–Schiemann
H_2N—⟨⟩—F (with Cl)	Halogen exchange
H_2N—⟨⟩—F (with F)	Halogen exchange
⟨⟩—F (with F)	Halogen exchange
Cl / N / CF_3 pyridine	Side-chain chlorination/fluorination
F—⟨⟩—C(=O)—⟨⟩—F	Balz–Schiemann
Br—⟨⟩—CH_2Br (with F)	Balz–Schiemann
⟨⟩—F	Balz–Schiemann
F—⟨⟩—F (with F)	Halogen exchange
⟨⟩—CF_3 (with NH_2)	Side-chain chlorination/fluorination
H_2N—⟨⟩—OCF_3	Side-chain chlorination/fluorination
⟨⟩—⟨⟩—F (with F)	Halogen exchange

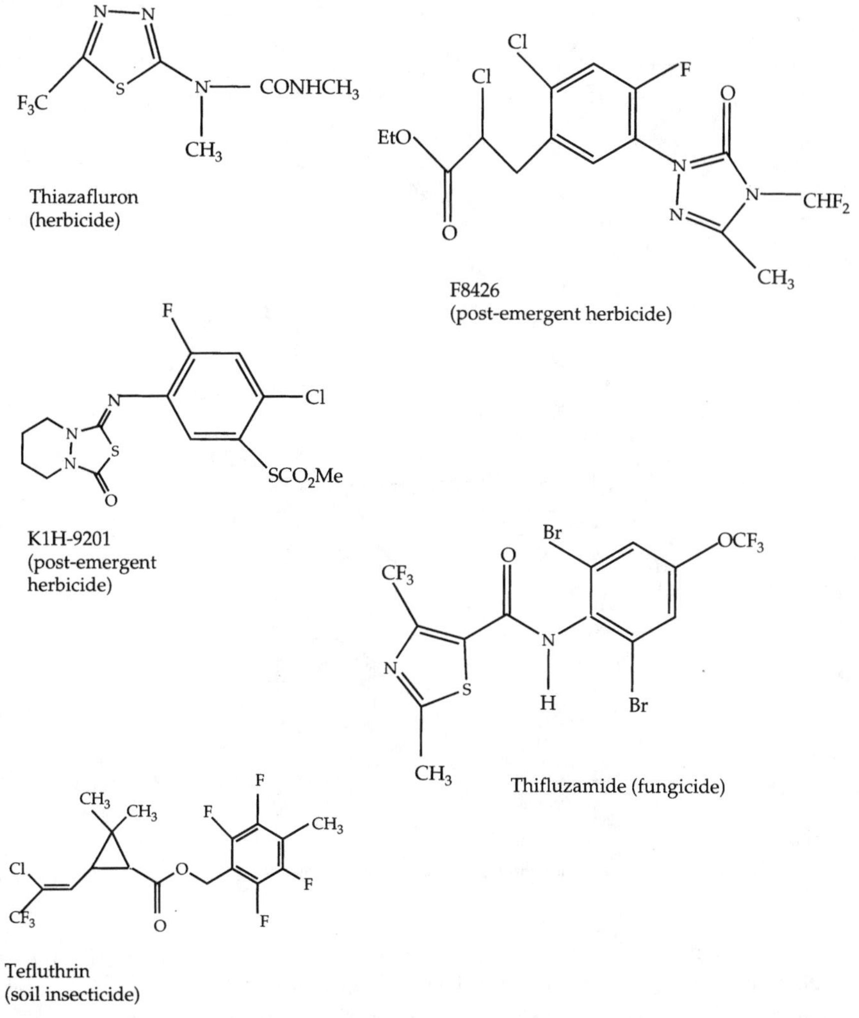

FIGURE 1.3 Examples of fluorine-containing pesticides.

terminal amino acid groups in proteins. The use of pentafluorophenol enables the easy formation of pentafluorophenoxy esters, which are of great value in peptide synthesis. Two important factors lead to the importance of fluorine in dyes:

1. Fluorine can act as a nucleofugal group in the process of fixing the dye to the fibers (for examples, see Figure 1.4)[29]
2. Fluorine can increase the reactivity of the dye towards the nucleophilic groups of the fibers, without itself being the leaving group, for example in fluoropyrimidines[24]

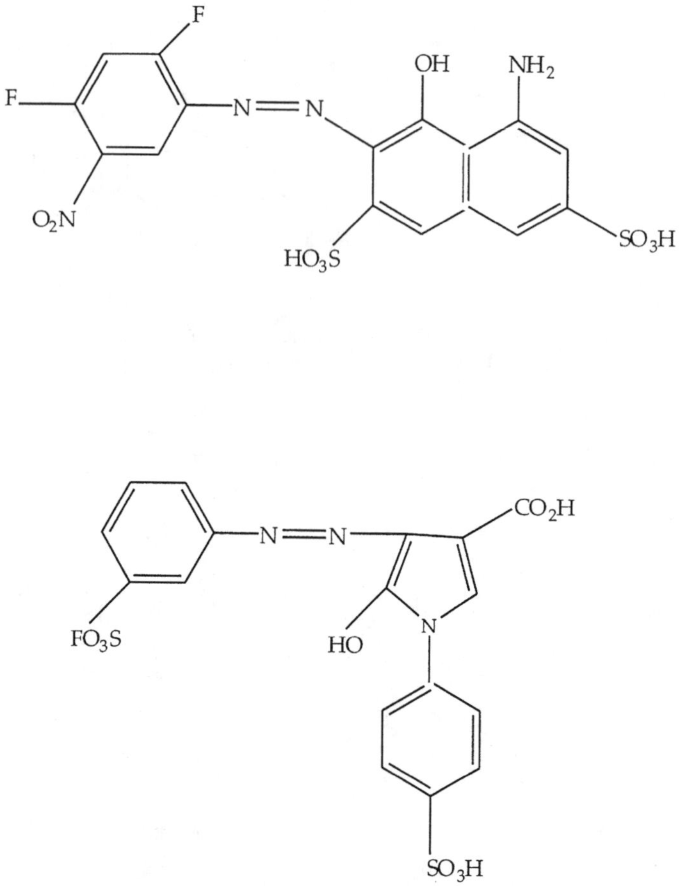

FIGURE 1.4 Examples of fluorine-containing dyes in which the fluorine acts as a leaving group in dye fixing.

In the area of advanced materials, fluorine has had a great impact on the development of liquid crystals and a large proportion of the new liquid crystal products are based on aromatics possessing small fluorine-containing substituents, such as F and CF_3. The introduction of just one or two fluorines into the molecule can have marked effects on the dielectric anisotropy. In some cases, this can be used to create a large negative dielectric anisotropy (suitable for ferroelectric liquid crystal displays) without the large increase in viscosity that can accompany substitution with other effective, but larger, groups (notably CN) (Figure 1.5). A potentially exciting area of application for the future is the use of fluorine substituents to usefully alter the properties of polyaromatics, polymers that will play an important role in the emergence of advanced materials. There has been a rapid growth of interest in this subject in the last 5 years with interesting new materials, including CF_3-containing polyimides,[30] CF_3-containing polamides,[31] CF_3-containing poly(aryl ether oxazole)s,[32] CF_3-containing polyarylsulfones,[33] and poly(ether ether ketone)s.[34] The

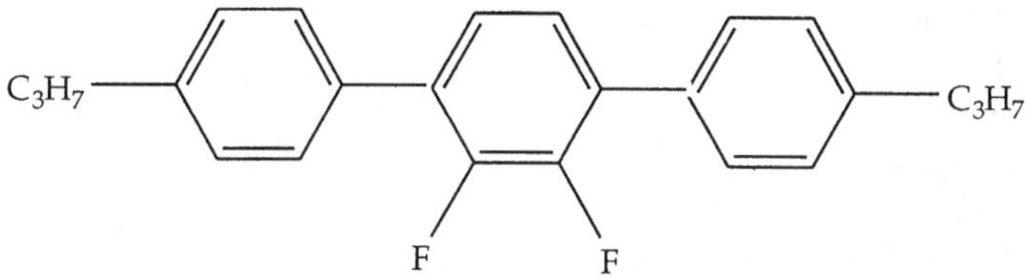

FIGURE 1.5 Example of a fluorine-containing ferroelectric liquid crystal.

fluorine substituents can affect the polymer T_g, thermal stability, solubility, and processability, and perhaps most significantly, the hydrophobicity of the material, even at quite low fluorine levels.[33,34] Applications in the electronic circuitry and aerospace industries are likely, especially if costs can be kept reasonable by the use of low-fluorine polymers and/or thin coatings.

1.4 RAW MATERIALS

The world's main reserves of fluorine are in the ores fluorospar and phosphate rock. These ores contain the minerals fluorite, CaF_2, and fluoroapatite, $Ca_5(PO_4)_3F$. There are more than 100 minerals containing fluorine as a result of the similarity in sizes of F^- and O^{2-} which enables fluorine to replace oxygen geochemically. Fluorine is not a rare element and is actually the 13th most abundant element in the Earth's crystal rocks, significantly greater than chlorine. The situation is reversed in seawater, where chlorine is almost 10^6 more abundant than fluorine. The estimated total quantity of fluorine in the whole Earth is some 10^{20} Kg, so that future supplies of the element would seem secure.

Natural products containing fluorine are rare and almost nothing is known about their origins. There are actually 10 identified naturally occurring organofluorine metabolites but none of these are fluoroaromatics.[35]

1.5 LABORATORY AND INDUSTRIAL METHODS OF FLUORINATION[3,36-40]

The preparation of highly fluorinated compounds generally involves the use of one of three fluorination methods based on:

1. Elementary fluorine (including the use of F_2 diluted with an inert gas)
2. High oxidation state metal fluorides (e.g., CoF_3)
3. Electrochemical fluorination (typically using HF containing an ionic fluoride)

The preparation of low-fluorinated or selectively fluorinated compounds can involve a much broader range of fluorinating agents and can involve a much wider choice of organic substrates. Thus, reaction types in this category that are relevant to the synthesis of fluoroaromatics include:

Cl, Br, I $\rightarrow$ F, CF_3, SCF_3, etc.

H $\rightarrow$ F

OH $\rightarrow$ F

NH_2 $\rightarrow$ F

NO_2 $\rightarrow$ F

CO_2H $\rightarrow$ CF_3

Fluorinating agents can be subclassified as (1) nucleophilic fluorine transfer reagents, (2) electrophilic fluorine transfer reagents, and (3) radical fluorine transfer reagents, although many reagents are not that easily placed in only one category.

Nucleophilic fluorine transfer reagents include metal fluorides, MF_n, hydrogen fluoride and its complexes, HF, $py(HF)_x$, $R_4NF(HF)_x$, etc., tetrafluoroboric acid, HBF_4, sulfur tetrafluoride and analogues, SF_4, $(Et_2N)SF_3$, onium fluorides R_4NF and R_4PF, metal perfluoroalkyls, including $CuCF_3$ and $CuSCF_3$, and halogen fluorides XF_n.

Metal fluorides are widely used as sources of F^-, and potassium fluoride is the most popular reagent in this category, being inexpensive if not especially reactive (due to its high lattice energy and poor solubility in all but highly protic solvents that are unsuitable for fluorination reactions). The more soluble and more reactive onium fluorides can be used for more difficult reactions or where milder conditions are desirable, but their cost and poor thermal stability limits their value. Catalysts, notably phase transfer catalysts, are widely used in fluoride reaction systems.

Metal (and onium) fluorides are most commonly used for the substitution of halogens directly bonded to the ring (i.e., ArCl $\rightarrow$ ArF). Aromatic nitro groups are attractive alternative leaving groups (ArNO$_2$ $\rightarrow$ ArF) and several recent developments in fluoride systems have helped to make fluorodenitration a viable alternative to halogen exchange, although industrial processes are very largely in the latter category. Multiple substitution of halogens from groups attached to the ring (e.g., $ArCX_3$ $\rightarrow$ $ArCF_3$; $ArSCX_3$ $\rightarrow$ $ArSCF_3$) normally requires acid catalysis (to weaken the C-X bond, since attack by F^- at the sterically hindered carbon is hindered) and hydrogen fluoride or HF-based reagents are normally favored in such reactions. Alternative routes to CF_3 substituted aromatics based on sulfur tetrafluoride ($ArCO_2H$ $\rightarrow$ $ArCF_3$) are useful laboratory methods but the expense of the reagent precludes large-scale use. HF can also be used to replace amino groups ($ArNH_2$ $\rightarrow$ ArF) although the better known method for this substitution is the "Balz-Schiemann" reaction employing BF_3. Substitution of halogen (commonly using a metal fluoride) and NH_2 (Balz–Schiemann) represent two of the most important industrial routes to selectively fluorinated aromatics, while HF-based industrial processes dominate in the manufacture of perfluoroalkyl (CF_3, SCF_3, OCF_3, etc.) substituted aromatics.

Nucleophilic sources of perfluoroalkyl groups such as CF_3 and CF_3S are becoming increasingly popular as more acceptable alternatives to HF-based technologies are sought. These are normally in the form of organometallics and the copper salts are the most popular, since they are not prohibitively expensive nor difficult to handle, and Cu^+ is particularly effective at helping to substitute unactivated aromatic halogens.

The traditional electrophilic fluorine transfer reagents, fluorine, perchloryl fluoride ($FClO_3$), and trifluoromethyl hypofluorite (CF_3COOF) can be considered to be vigorous fluorinating agents — they are highly reactive, unstable, and toxic and require careful handling using specialized equipment.[36,37] Improvements in the selectivity of reaction and in the handling of elemental fluorine have been achieved through the use of diluent gases such as nitrogen, but reagents in this class have little value in aromatic fluorine chemistry.

Cesium fluoroxysulfate, $CsSO_4F$, is a solid electrophilic fluorinating agent that has been available since 1979[41] and, while it offers some advantages over the more traditional electrophilic fluorinating agents, it is a hazard and can explode under certain conditions.[37] The reagent has been used in the fluorination of aromatics, including important heterocycles such as uracil.[42]

N-F electrophilic fluorine transfer reagents were developed to overcome the problems associated with earlier reagents in this class, i.e., instability, toxicity, lack of selectivity, and difficulty in handling.[37,38] They are mostly, but not all, based on cyclic nitrogen compounds where the nitrogen is generally in the form of a N-F bond. Substituents on the ring can help to control solubility and reactivity (Figure 1.6). The better reagents in this category are stable, high melting solids with good solubility in a variety of solvents. The more active reagents can directly fluorinate aromatic rings, although forcing conditions will be required for unactivated systems. The less reactive reagents are capable of fluorinating Grignards and other metal salts.

Electrophilic reagents are commonly used in the preparation of some perfluoroalkyl-substituted aromatics. CF_3SCl (in the preparation of $ArSCF_3$) and $(CF_3SO_2)_2O$ (in the preparation of ArO_2SCF_3) are among them. Even trifluoromethylaromatics can be prepared via the electrophilic route by using S-, Se-, and Te-trifluoromethylated dibenzoheterocyclic onium salts as sources of CF_3^+ (see Chapter 5).

While fluorine is capable of acting as a reasonably controlled electrophilic fluorinating agent, its behavior is more commonly associated with a highly unselective radical fluorinating agent.[36] In this capacity it has no value in selective aromatic fluorination. Other reagents that can and often do act via radical or radical-anion pathways include the noble gas fluorides (although again here, careful control of conditions can lead to useful selectivity even in aromatic fluorination; extensive studies have been carried out on XeF_2 in particular[37] and photofluorinations using OF compounds such as CF_3OF).[36]

Radical routes can be used in the synthesis of perfluoroalky-substituted aromatics. The CF_3 radical (e.g., from photochemical activation of CF_3I) can react with aryl thiol salts to give the SCF_3-substituted products and directly with electron-rich aromatic substrates to give trifluoromethylaromatics.

The various methods of fluorination and perfluoroalkylation are described in more detail in Chapters 2 to 7.

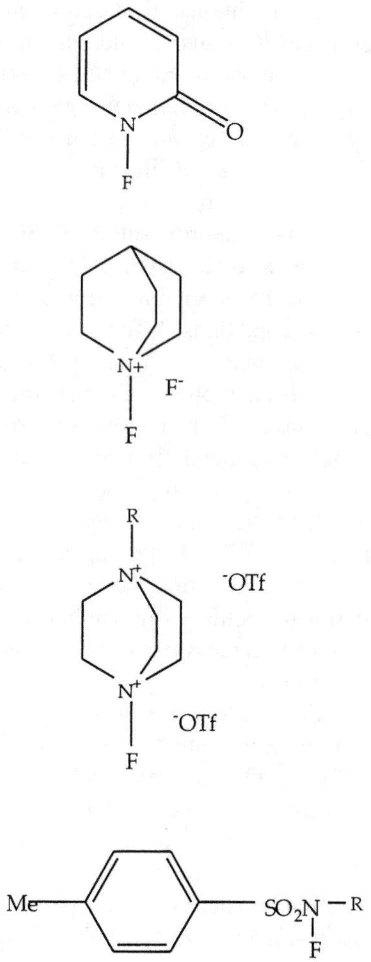

FIGURE 1.6   Some N-F electrophilic fluorine transfer reagents.

REFERENCES

1. R. Schmitt and H. von Gehren, *J. Prakt. Chem.,* 1870, 1, 394.
2. E. Paternò and V. Oliveri, *Gazzetta,* 1883, 13, 533.
3. *Fluorine: The First Hundred Years (1886–1986),* eds. R.E. Banks, D.W. A. Sharp and J.C. Tatlow, Elsevier, Lausanne, 1986.
4. H. Bart, German Patent, 281 055 (1913).
5. G. Balz and G. Schiemann, *Chem. Ber.,* 1927, 60, 1186.
6. G. Schiemann, *Chem. Ztg.,* 1930, 54, 269.
7. A. Roe, *Org. React.,* 1949, 5, 193.
8. H. Suschitzky, *Adv. Fluorine Chem.,* 1965, 4, 1.
9. G. Schiemann and B. Cornils, *Chemie und Technologie cyclische Fluorverbindungen,* Enke, Stuttgart, 1969.
10. H.B. Gottlieb, *J. Am. Chem. Soc.,* 1936, 58, 532.
11. A.E. Chichibabin and M.D. Rjazancev, *J. Russ. Phys. Chem. Soc.,* 1915, 46, 1571 (*Chem. Abstr.,* 1916, 10, 2898).
12. E.M. McBee, V.V. Lindgren and W.B. Ligett, *Ind. Eng. Chem.,* 1947, 39, 378.
13. J. Desirant, *Bull. Soc. Chim. Belg.,* 1958, 67, 676.
14. W. Prescott, *Chem. Ind. (London),* 1978, 56.
15. W.A. Sheppard, *J. Am. Chem. Soc.,* 1965, 87, 2410.
16. R.D. Chambers, *Fluorine in Organic Chemistry,* Wiley, New York, 1973.
17. J.D. Roberts, R.L. Webb and E.A. McElhill, *J. Am. Chem. Soc.,* 1950, 72, 408.
18. F.S. Fawcett and W.A. Sheppard, *J. Am. Chem. Soc.,* 1965, 87, 4341.
19. D.T. Clark, N.J. Murrell and J.M. Tedder, *J. Chem. Soc.,* 1963, 1250.
20. M.J.S. Dewar and T.G. Squires, *J. Am. Chem. Soc.,* 1968, 90, 210.
21. B.E. Smart in *Organofluorine Chemistry: Principles and Commercial Applications,* eds., R.E. Banks, B.E. Smart and J.C. Tatlow, Plenum Press, New York, 1994, pp. 57–88.
22. A.L. Brown, J. Chiang, A.J. Kresge., Y.S. Tang, and W.-H. Wang., *J. Am. Chem. Soc.,* 1989, 111, 4918.
23. P.N. Edwards in *Organofluorine Chemistry: Principles and Commercial Applications,* eds. R.E. Banks, B.E. Smart and J.C. Tatlow, Plenum Press, New York, 1994, pp. 501–541.
24. *Organofluorine Chemicals and Their Industrial Applications,* ed. R.E. Banks, Horwood, Chichester, 1979.
25. *Preparation, Properties and Industrial Applications of Organofluorine Compounds,* ed. R.E. Banks, Ellis Horwood, Chichester, 1982.
26. *Organofluorine Chemistry: Principles and Commercial Applications,* eds. R.E. Banks, B.E. Smart and J.C. Tatlow, Plenum Press, New York, 1994.
27. F. Coppola, *Gazzetta,* 1883, 13, 521.
28. J.H. Fried and E.F. Sabo, *J. Am. Chem. Soc.,* 1953, 75, 2273.
29. A.H. Barrie, G. Booth and I. Durham, Br. Patent 882001 (1959), *Chem. Abstr.,* 1962, 56, 8962h.
30. T. Matsuura, N. Yamada, S. Nishi and Y. Hasuda, *Macromolecules,* 1993, 26, 419.
31. Z.G. Gardlund, *Polymer,* 1993, 34, 1850.
32. G. Maier, R. Hecht, O. Nuyken, K. Burger and B. Helmreich, *Macromolecules,* 1993, 26, 2583.
33. J.H. Clark and J.E. Denness, *Polymer,* 1994, 35, 2432.
34. J.H. Clark and J.E. Denness, *Polymer,* 1994, 35, 5124.
35. M. Meyer and D. O'Hagan, *Chem. Br.,* 1992, 785.
36. M.R.C. Gerstenberger and A. Haas, *Angew. Chem. Int. Ed. Engl.,* 1981, 20, 647.
37. J.A. Wilkinson, *Chem. Rev.,* 1992, 92, 505.
38. V. Murtagh, *Performance Chem.,* 1992, 27.
39. J. Mann, *Chem. Soc. Rev.,* 1987, 16, 381.
40. M.A. McClinton and D. McClinton, *Tetrahedron,* 1992, 48, 6555.
41. E.H. Appleman, L.J. Basil and R.C. Thompson, *J. Am. Chem. Soc.,* 1979, 101, 3384.
42. M. Zupar and S. Stavber, *J. Chem. Soc. Chem. Commun.,* 1981, 148.

Chapter 2

Halex Chemistry

2.1 INTRODUCTION

Swarts pioneered the use of metal fluorides as nucleophilic fluorinating agents and prepared many fluoroaliphatic compounds from their chlorinated analogues using the strong Lewis acidic antimony fluorides.[1] Many other metal and nonmetal fluorides have been used in aliphatic halex reactions, but the most commonly used fluorides in aromatic halex reactions are the alkali metal fluorides, and potassium fluoride in particular.

Aromatic halex was first described by Gottlieb[2] and developed extensively by Finger and Kruse.[3] Since then the reaction has been applied to the synthesis of many fluoroaromatic and fluoroheterocyclic compounds.[4] With regard to cost and availability, chloroaromatics are normally the substrates of choice and reactions with potassium fluoride (a compromise between fluoride activity and reagent cost) generally require chloroaromatics that are activated towards nucleophilic attack. Perhaps the best known reaction in this context is the fluorination of chloro-2,4-dinitrobenzene and this substrate can be used to exemplify the normal reaction mechanism that proceeds via a Meisenheimer (anionic) intermediate (Figure 2.1). These intermediates can actually be long lived enough to be observed spectroscopically, notably in the case of 1,3-dinitroaromatics and using the highly reactive and soluble tetrabutylammonium fluoride.[5] Dipolar aprotic solvents such as amides, sulfoxides, and sulfones, are generally preferred and, while it has been widely assumed that this is to enhance fluoride solubility, it is possible that their major role is actually to help stabilize Meisenheimer-type intermediates. The solubility of potassium fluoride in dipolar aprotic solvents is in the millimolar range and an interfacial mechanism (whereby substrate molecules react on the surface of the undissolved fluoride) may well compete with a normal solution mechanism.

Inductive and mesomeric effects must be considered in determining the likelihood of a substrate reacting in a halex reaction. Inductive effects decrease with distance from the point of nucleophilic attack and mesomeric effects are most powerful when exerted ortho or para. Chlorobenzene itself will react with KF but only under very forcing conditions (Figure 2.2), whereas both 2-chloronitrobenzene and 4-fluoronitrobenzene are sufficiently activated to enable attack under moderate conditions (<200°C).

Substrates that are activated only inductively can be susceptible to halex under moderate conditions but traditionally only in extreme cases. Thus, hexachlorobenzene will react with KF in a dipolar aprotic solvent at <200°C to give a mixture of chlorofluorobenzenes. This is not normally an effective route to the fully exchanged product, hexafluorobenzene, which is normally only a minor product although under

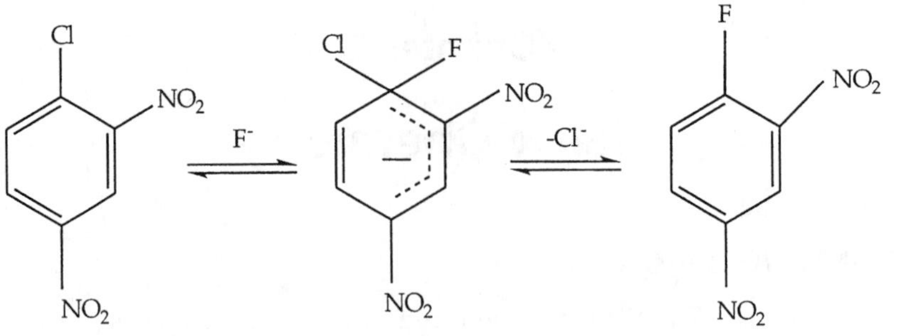

FIGURE 2.1   Mechanism for halex reactions.

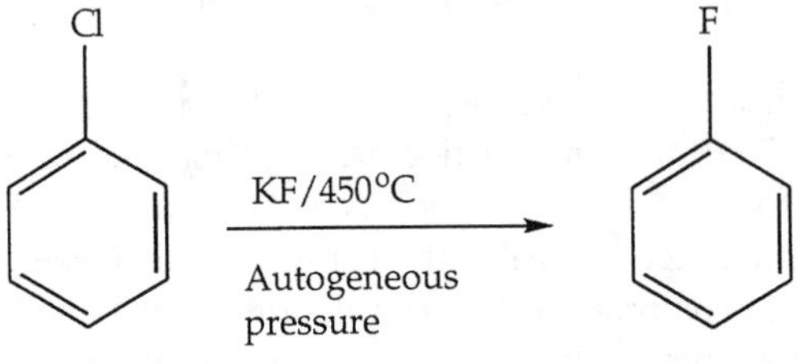

FIGURE 2.2   Halex reaction of chlorobenzene.

forcing (high-temperature) conditions it can be produced in yields of about 20% and this has formed the basis of a manufacturing process (hexafluorobenzene was once considered as a possible anesthetic). The major product is actually 1,3,5-trichloro-trifluorobenzene and this is a good example of the so-called Iπ effect.[6] This acts to inhibit substitution of groups para to fluorine due to the build up of charge in the anionic intermediate on the carbon adjacent to the electron-rich fluorine atom.

The Iπ effect can be used to give selectivity in halex reactions of multiply substituted substrates. More commonly, selectivity can result from different degrees of activation of halogens. Thus, 3,4-dichloronitrobenzene will react exclusively at the 4-position to give 3-chloro-4-fluoronitrobenzene (Figure 2.3). In contrast, 2,4-dichloronitrobenzene reacts at both chlorines giving a mixture of products with less than two mole equivalents of fluoride. With enough fluoride, the difluoro product can be obtained. The relative rates of substitution of the two chlorines in the latter substrate is not as might be expected. The displacement by fluoride of the *o*-chlorine proceeds at a higher rate than that of the *p*-chlorine, whereas the *p*-chlorine in 4-chloronitro-benzene reacts away much faster than the *o*-chlorine in 2-chloronitrobenzene.[3,7,8] These apparently contradictory results can be largely explained by two effects:

1.   The activating effect of the nitro group is predominately via a negative mesomeric effect and for this to be maximal the nitro group should be coplanar with the aromatic ring. Bulky atoms (e.g., chlorine) ortho to the nitro group can effectively twist the nitro out of the plane of the ring, thus reducing its mesomeric activating effect.

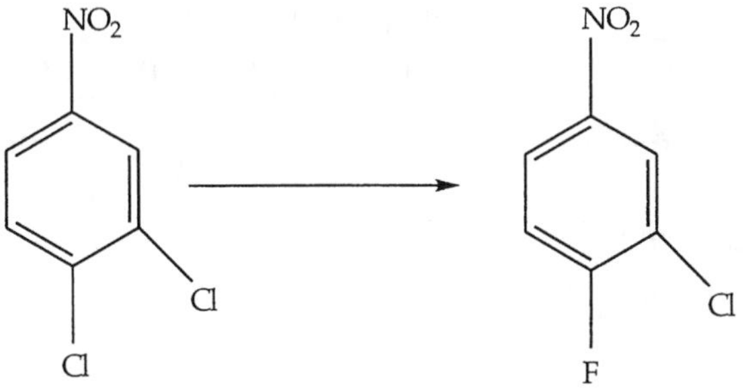

FIGURE 2.3 Selective halex reaction of 3,4-dichlorobenzene.

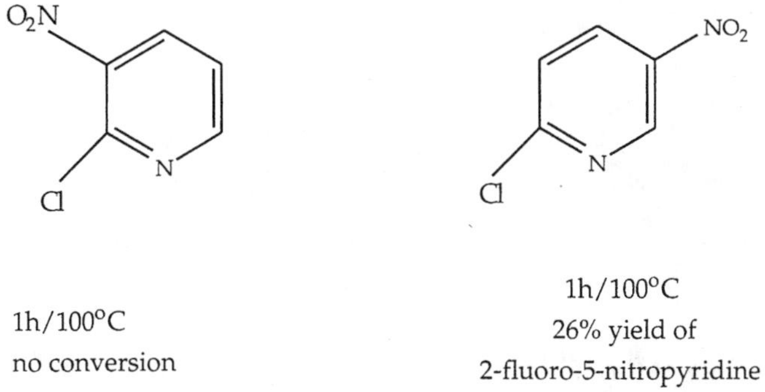

1h/100°C

no conversion

1h/100°C

26% yield of

2-fluoro-5-nitropyridine

FIGURE 2.4 Relative reactivities of chloronitropyridine towards potassium fluoride.

2. To a lesser extent the nitro group activates by inductive effects. These effects are proportional to distance so that the o-chlorine is significantly more activated than the p-chlorine.

The same factors can be used to explain the large difference in reactivities between 2-chloro-3-nitropyridine (unreactive under mild conditions) and its isomer, 2-chloro-5-nitropyridine (reasonably reactive under mild conditions) (Figure 2.4).[9]

The nitro group is the most powerful activating group of the more common aromatic substituents but other less activating substituents, including CN, CF_3, and the pyridine nitrogen, can be sufficient to promote halex reactions. Under moderate conditions a single one of these substituents will not be sufficient to activate a chlorine towards halex, but effects are additive, so that the presence of two or more of these groups should enable reaction to occur. This is nicely illustrated by the relative reactivities of a series of chloropyridines (Table 2.1).[10,11]

While NO_2 is the activating group par excellence in halex chemistry, it can also act itself as a leaving group, sometimes in competition with halex. Fluorodenitration has been known for over 40 years and was reported in the pioneering work of Finger and Kruse as a side reaction,[3] but it is only in recent years that it has been considered

Table 2.1 Relative Reactivities of Chloropyridines in Halex Reactions

Substrate	Conditions	Yield of Difluoro Product
	1.5 hr/200°C	83–94%
	7 hr/200°C	72%
	200 hr/200°C	52%
	328 hr/200°C	33%

as a viable route to fluoroaromatics. Generally, the nitro group can be expected to become a better leaving group when it loses coplanarity with the ring. Thus, bulky substituents ortho to the nitro group, as well as the normal activating effects for nucleophilic aromatic substitution, are relevant.

The relative paucity of reported examples of fluorodenitration reactions compared to halex reactions despite the obvious economic appeal of nitroaromatics as starting materials (as witnessed by the great importance of the indirect conversion of nitroaromatics to fluoroaromatics via anilines in Balz-Schiemann-type reactions) is due in part to the relatively high costs of the suitably activated substrates. Thus, in a simple example (Figure 2.5), the costly fluorodenitration substrate (**1**) is not a viable alternative to the relatively inexpensive halex substrate (**2**) for the preparation of 4-fluoronitrobenzene (**3**). In many cases the more common meta isomer of nitroaromatics possessing another strongly activating substituent naturally makes their reactivity towards nucleophilic substitution rather low. The appeal of fluorodenitration as a synthetic route will often occur when the corresponding chloro and nitro meta-substituted aromatics are compared as substrates for preparing metafluoroaromatics. Thus, 1,3-dinitrobenzene (**4**, Figure 2.5) is a much more likely, if still very

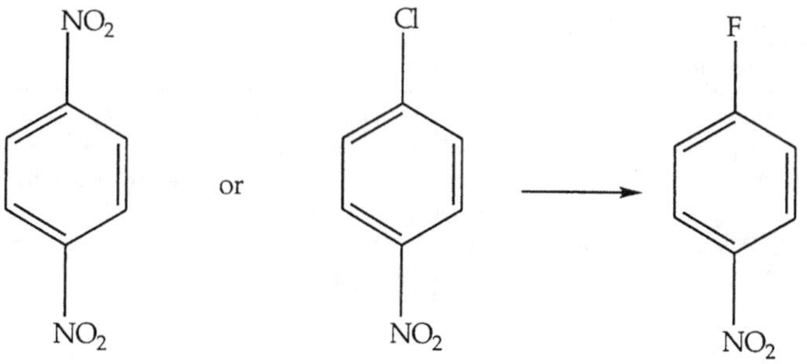

FIGURE 2.5 Substrates for the preparation of fluoroaromatics.

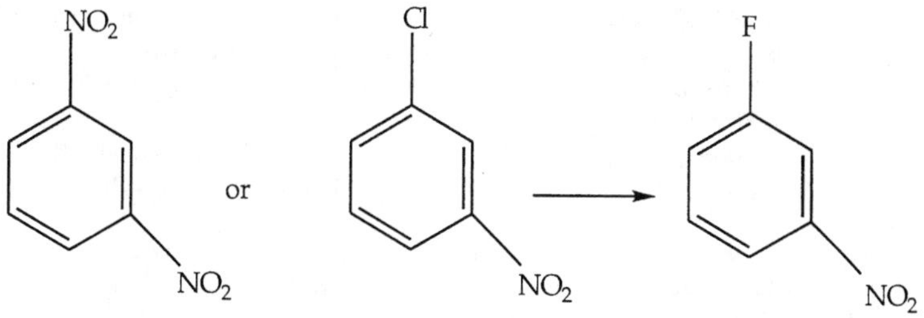

difficult, substrate for the preparation of 3-fluoronitrobenzene (**6**) than 3-chloronitrobenzene (**5**), because of the larger inductive effect of NO_2 compared to chlorine. Difficulties in fluorodenitration reactions can still remain and these are discussed in Section 2.5.2.

2.2 FLUORIDE REAGENTS

An enormous number of fluorides are known. Most elements, including the noble gases combine with fluorine which is also capable of inducing elements into unusually high oxidation states, a phenomenon known as hypervalency. Fluorides find applications in many areas of organofluorine chemistry, depending upon their chemical properties. They are commonly divided into the categories of nucleophilic fluorinating agents and electrophilic fluorinating agents. The majority of fluorides are nucleophilic in nature, but it is only those based on a hard cationic center that have sufficient F^- character to make them suitable for halex-type reactions (Table 2.2).

Table 2.2 Fluoride Reagents for Aromatic Halex and Related Reactions

Reagent	Availability	Comments
KF	Readily available and inexpensive	Most commonly used source of nucleophilic fluorine in the laboratory and in manufacturing processes; very low solubility in all but protic solvents (strong hydrogen bonding)
CsF	Readily available but expensive; sometimes used in combination with KF	More soluble and more reactive than KF; very hygroscopic but can be made anhydrous
NaF	Readily available and inexpensive	Less soluble and less reactive than KF
Bu$_4$NF	Available as hydrate or as a solution; expensive; can be prepared from aqueous hydroxide (HF) or by metathetical reactions	One of the most reactive sources of F$^-$ but effectively impossible to dry completely; thermally unstable (decomposes at ca. > 100°C); soluble in polar aprotic solvents but also likely to attack such solvents (MeCN, DMSO, THF), especially when there are no protic molecules available
Et$_4$NF	Available as hydrate; expensive; can be prepared from inexpensive chloride	Less reactive than heavier ammonium fluorides and soluble only in more polar solvents (e.g., MeCN); can be made anhydrous; hygroscopic but relatively easy to handle
Ph$_4$PHF$_2$ and other bifluorides	Can be prepared from hydroxide (HF)	Less hygroscopic than the fluorides and mostly good solubilities but generally significantly less reactive
ZnF$_2$	Can be made from the metal, oxide, bromide, or sulfide	Highly toxic and infrequently used
KF/18-crown-6	Commercially available	More reactive than KF itself due to increased solubility but increase rarely outweighs added costs and toxicity problems due to crown ether
KF/phase transfer catalyst	Many reagents are readily available	Popular combination for many halex-type reactions; high-temperature reactions limited to Me$_4$NCl or Ph$_4$PBr (typically) due to low thermal stability of many onion compounds

Calcium fluoride is the major source of fluorine and certainly fits into the category of a hard cationic-centered fluoride. Its use in nucleophilic fluorinations would have enormous economic advantage but, unfortunately, the extremely high lattice energy of the salt renders it insoluble and unreactive so that its use is restricted to being the major source of HF via reaction with hot, concentrated sulfuric acid. Generally, only monovalent fluorides have a sufficiently low lattice energy to make them reactive enough for halex-type applications. Of these the alkali metal fluorides have traditionally been the most popular sources of F$^-$ with solubility and reactivity following the trend in reducing lattice energies:

LiF < NaF < KF < RbF < CsF

In practice, KF is widely used, being a compromise between reactivity and cost. The less reactive NaF may be a viable alternative for highly reactive substrates and the more reactive (but also significantly more expensive) CsF is also sometimes used in difficult reactions. Interestingly, KF and CsF can be used in combination whereby the overall fluoride activity in a halex-type reaction is greater than KF alone but less than that of CsF alone, suggesting a catalytic effect by the cesium cation.

It is important to note that KF-based halex reactions are generally energetically favorable, e.g.,

$$C_6Cl_{6\ (g)} + 6KF_{(s)} \rightarrow C_6F_{6(g)} + 6KCl_{(s)}\ \Delta H = -125\ kJ\ mol^{-1}$$

The same reaction involving NaF is actually unfavorable.[12] The rather slow rates of reaction and the need for elevated temperatures normally associated with KF halex reactions are largely a result of poor availability of the fluoride anion (the rate-determining step in most halex-type reactions is the reaction of the free fluoride with the substrate to produce the anionic [Meisenheimer] intermediate). This is considered to be a result of the very limited solubility of KF (and indeed all metal fluorides).

Potassium fluoride only enjoys reasonable solubility in protic solvents that are powerful hydrogen bond electron acceptors (proton donors). These include H_2O, $CH_3CONH(CH_3)$ and the lighter carboxylic acids. This can be attributed to the powerful hydrogen bond electron donor (proton acceptor) properties of F^-. Thus, in the case of acetic acid, for example, KF is soluble up to about 4 M due to the formation of very strong (>100 kJ mol^{-1}) hydrogen bonds (Figure 2.6).[13] Without the formation of such energetically favorable interactions, the lattice energy of KF cannot be overcome. Unfortunately, the result of such strong hydrogen bonding is that the charge and, hence, nucleophilicity of the fluoride is significantly reduced and only the most electrophilic of substrates are liable to halex chemistry in KF/CH_3CO_2H (or any other fluoride/protic solvent system). Other substrates are more likely to undergo reactions with the protic molecule, the nucleophilicity of which is enhanced by hydrogen bonding (Figure 2.7).[14] This is a powerful reminder of the basicity of F^- and the need to keep nucleophilic fluorine transfer (e.g., halex) systems free from protic molecules or indeed any molecules liable to attack by base.

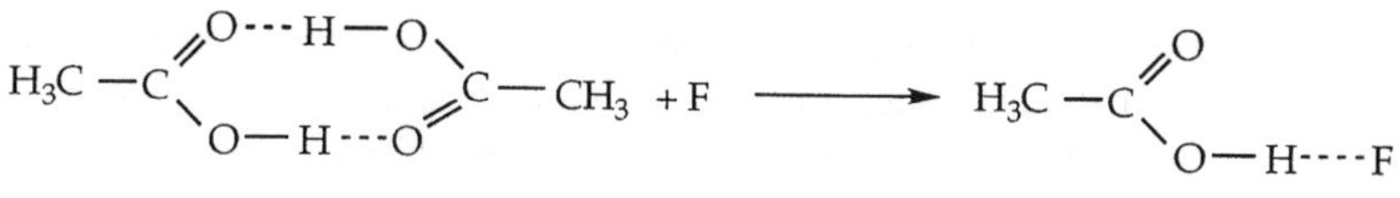

FIGURE 2.6 Formation of strong hydrogen bonds between acetic acid and F^-.

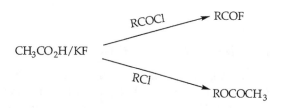

FIGURE 2.7 Reactions of KF/CH_3CO_2H with electrophiles.

It is necessary, therefore, to run almost all halex reactions with fluorides in aprotic solvents. Generally, dipolar aprotic solvents are preferred. These solvents are thermally stable and generally encourage nucleophilic substitution reactions on activated aromatic substrates by stabilization of Meisenheimer intermediates (Figure 2.1). It is often said that these solvents are used to enhance KF solubility, but even at close to reflux temperatures, KF solubilities in dipolar aprotic solvents are only at a millimole level. It is possible that such reactions involve an interfacial mechanism where much of the reaction occurs at the KF surface. It is known that the surface area of the fluoride and other factors such as the stirring rate can significantly affect rate of reaction and these observations would be consistent with at least a contribution from an heterogeneous reaction. More soluble fluorides such as CsF should contribute more to a homogeneous reaction.

Numerous attempts have been made to enhance the reactivity of KF in particular, in nucleophilic fluorine transfer reactions. The more effective general methods are briefly described below.

2.2.1 Surface Area

If an interfacial mechanism is relevant in nucleophilic fluorinations or if the rate of reaction is limited by dissolution of fresh KF into solution, the surface area of the salt is clearly important. In this context, spray-dried KF (surface area 1.3 m^2 g^{-1}) is now widely regarded as being superior to ordinary (calcined) KF (surface area 0.1 m^2 g^{-1}).[15]

A popular method of enhancing the activity of salts as reagents by increasing their effective surface area is via so-called supported reagents.[16] Unfortunately, attempts to improve the reactivity of KF as a source of nucleophilic fluorine via this methodology have been largely unsuccessful. Potassium fluoride supported on alumina, for example, is a remarkably effective base but is unreactive in halex reactions due to surface OHF-hydrogen bonding and other surface chemistry (such as reaction to form the strong Al-F bonds).[17] Few likely support materials will be resistant to attack by fluoride. Calcium fluoride is available in a high surface area form (via preparation by precipitation) although the surface area (10 to 20m^2 g^{-1}) is an order of magnitude less than the surface areas of common chromatographic supports such as alumina and silica. KF-CaF$_2$ has been shown to be an effective reagent in aliphatic halex reactions, although improvements over spray-dried KF are rather small.[18,19]

New effective methods of significantly increasing the surface area of KF or NaF are likely to have great significance to the economic growth of aromatic fluorine chemistry.

2.2.2 Phase Transfer Catalysis and Related Methods

One of the most effective general methods of enhancing the rate of nucleophilic fluorine transfer reactions based on insoluble fluorides such as KF is solid-liquid phase transfer catalysis (SLPTC). The extremely strong hydration experienced by fluoride effectively rules out the alternative liquid-liquid phase transfer catalysis involving an aqueous solution of the fluoride. Transfer of F⁻ from solid to liquid via formation of an onium fluoride will enhance the homogeneous reaction due to increased fluoride concentration in solution. It is also possible that a phase transfer catalyst can assist heterogeneous reaction by "loosening" surface fluoride or possibly by helping to stabilize reaction intermediates.

The most widely used phase transfer catalysts in aromatic fluorinations are tetramethylammonium chloride and tetraphenylphosphonium chloride (or bromide). Alkyl groups larger than methyl are prone to base-catalyzed decomposition and this rules out the use of the large majority of onium compounds, at least at the temperatures of most nucleophilic aromatic fluorinations (>100°C).

Several commercial processes based on the use of tetramethylammonium chloride have been successfully developed to run at full manufacturing scale. Tetraphenylphosphonium chloride and bromide are especially stable to heat, and higher-temperature (>200°C) processes based on their use have been developed. The drawbacks with the phosphonium salts are their high cost and the risk of loss of the catalyst in wet systems due to formation of the highly stable triphenylphosphine oxide from base hydrolysis.[20,21]

Crown ethers have proven to be relatively ineffective catalysts in nucleophilic fluorinations possibly due to K(crown)⁺.... F⁻ interactions in solution.[22]

Various other phase transfer-based systems have been described. Combinations of crown-type compounds (crown ethers or capped polyethylene glycols) and onium salts can be effective, especially in more demanding reactions.[23] Longer-chain tetraalkylammonium compounds have been claimed as effective catalysts in aromatic halex and, less commonly, Lewis acids have been reported as assisting halex reactions, presumably through weakening of the ring-chlorine bond although the combination of a Lewis base (fluoride) and a Lewis acid is clearly problematic.[24]

2.2.3 Soluble Fluorides

An alternative approach is to use more soluble sources of F⁻,[25] such as R_4NF (e.g., Bu_4NF,[26] Me_4NF[27]), R_4NHF_2 (e.g., Bu_4NHF_2[28]), and R_4PHF_2 (e.g., Ph_4PHF_2[29]). These are considerably more reactive than metal fluorides such as KF, largely due to their excellent solubility in polar aprotic solvents. Unfortunately, the most reactive of the onium fluorides are subject to decomposition due to fluoride-catalyzed Hoffmann-type elimination of alkyl group. Thus, the highly reactive tetrabutylammonium fluoride (TBAF) will decompose, giving off butene at temperatures above 50°C (Figure 2.8). Tetramethylammonium fluoride (TMAF) is stable even at temperatures above 100°C as a result of the lack of β-protons, although it is significantly less reactive than TBAF.[27] Tetraphenylphosphonium bifluoride is also quite stable to high temperatures.[30]

Reactivity of the onium fluorides also roughly parallels hygroscopicity. TBAF, for example, has never been prepared in an anhydrous state. The best attempts to remove water have reduced it to less than one mole equivalent, at which point the reagent is a highly reactive "oil" which will attack many solvents.[26,31] The only organic ammonium fluoride that has been prepared in a truly anhydrous form is

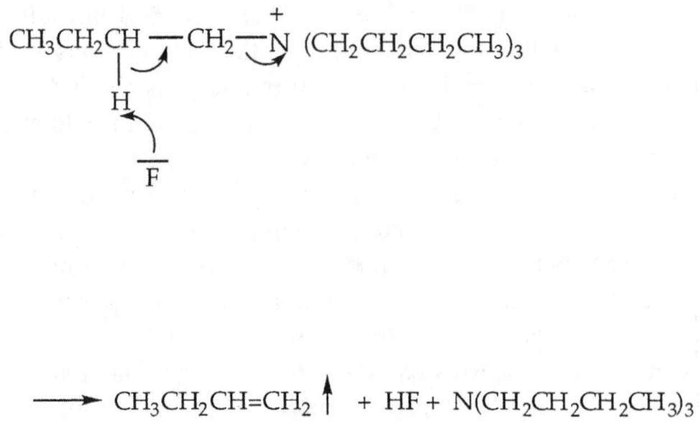

$$CH_3CH_2CH \overset{|}{\underset{H}{}}\!-\!CH_2\!-\!\overset{+}{N}(CH_2CH_2CH_2CH_3)_3$$

$$\longrightarrow CH_3CH_2CH{=}CH_2 \uparrow \; + \; HF \; + \; N(CH_2CH_2CH_2CH_3)_3$$

FIGURE 2.8   Decomposition of tetrabutylammonium fluroide.

TMAF, although the drying procedure is extremely tedious and the product is extremely hygroscopic.[32] Fortunately, *in situ* azeotropic drying is an effective method for TMAF activation.[27]

Cost is also a drawback to the use of soluble fluorides and is likely to restrict their use to reactions of particularly difficult substrates. Specialist uses have been developed, with TMAF being a particularly effective reagent for fluorodenitrations (see Section 2.5.2).

2.3 PHYSICAL PROPERTIES OF FLUOROAROMATICS

The introduction of a single fluorine into an aromatic nucleus can have significant effects on the physical properties of that molecule. Thus, while the boiling point of fluorobenzene (85°C) is very close to that of benzene (80°C), the melting points differ by some 48°C. Interestingly, the boiling and melting points of hexafluorobenzene are actually closer to benzene than to fluorobenzene, although this does not apply to other physical properties. Various physical properties of fluorobenzene, benzene, and hexafluorobenzene are compared in Table 2.3.

Table 2.3   Some Physical Properties of Benzene, Fluorobenzene, and Hexafluorobenzene

	Benzene	Fluorobenzene	Hexafluorobenzene
m.p./°C	5.5	−42	4
b.p./°C	80	85	81
d/gcm^{-3}	0.874	1. 024	1.612

The electron withdrawing ability of fluorine can be nicely illustrated by looking at the effects of ring fluorination on the acidity of phenols and carboxylic acids (Table 2.4). The effects are significant but not great and correlate with the distance of the fluorine from the ionizable group. Interestingly, the effects can be much greater, as in the case of imidazole, where the introduction of one fluorine substituent can increase the acidity by almost 5 pKa units (Figure 2.9).

Table 2.4 Effect of Fluorine and Other Substituents on the Acidity of Phenols and Benzoic Acids

| | | pK$_a$ Values | |
		X = OH	X = CO$_2$H
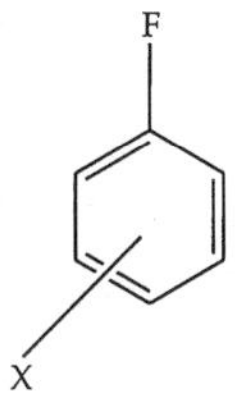	no F	10.00	4.19
	ortho-F	8.70	3.27
	meta-F	9.21	3.86
	para-F	9.91	4.14

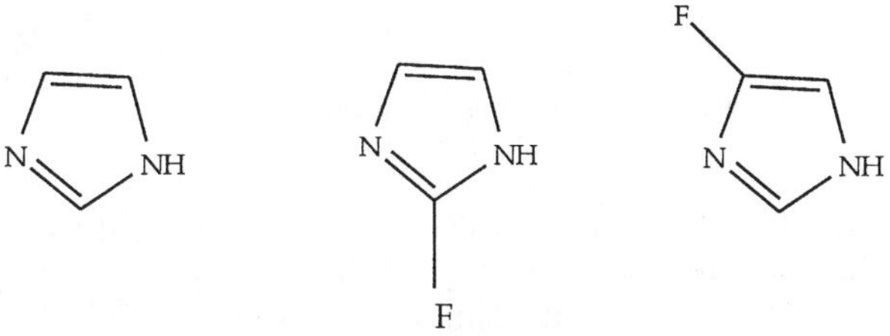

pKa = 7.08   pKa = 2.40   pKa = 2.44

FIGURE 2.9 Effects of fluorine on the acidity of imidazole.

Fluorine incorporation into the aromatic components of drugs has been enormously important in the improvement of efficiency of many pharmaceuticals.[33] One very important property of a drug is its rate of absorption into different environments. Biological membranes consist of lipid bilayers and represent one of the major hurdles that a drug molecule needs to surmount if it is to reach the target site (cell penetration, blood-organ, gut-blood, etc.). For an ionizable drug, the rate of absorption is determined by several physicochemical properties:

- pK$_a$
- Molecular size
- Lipophilicity
- Solubility

Fluorine substitution is described as being "size-neutral" so that replacing one of the hydrogens in benzene by fluorine for example, has no significant effect on the molecular size, whereas replacement by other groups such as chlorine, methyl, and trifluoromethyl have large effects on molecular size.

Lipophilicity is slightly increased on going from benzene to fluorobenzene, although the effect is small compared to that caused by chlorine and trifluoromethyl (see Chapter 1). The combined benefits of a small increase in lipophilicity and rather larger effects on pK$_a$ can have profound effects on drug efficacy. Thus, almost all of the clinically useful quinalone antibacterial agents contain a fluorine atom at the aromatic C-6 position, largely because there is an increase of up to 70 times in cell penetration.[34]

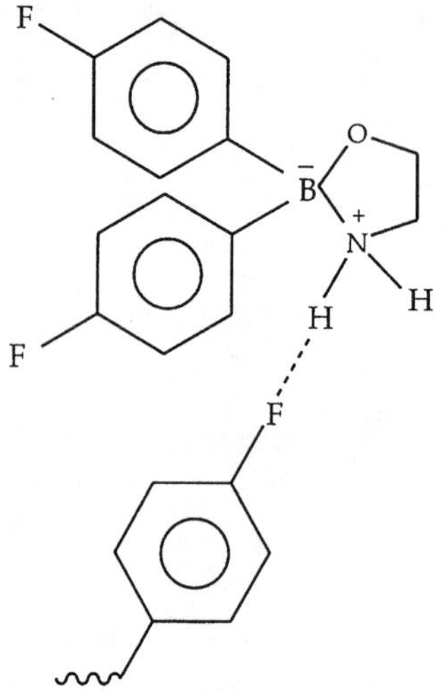

FIGURE 2.10   A rare example of an aromatic C-F hydrogen bond.

Fluorine substitution on an aromatic system will typically cause a change in the affinity of the molecule towards biological macromolecules (most often proteins) by a factor of 1 to 20. Since it is this affinity that is at the heart of drug potency it is important to try to understand the properties of the fluorinated molecule that result in such changes, so that these can be optimized.

Since the hydrogen bonding properties of the fluoride ion are so pronounced and hydrogen bonding can play such an important role in biochemical processes, it does not seem unreasonable to expect aromatic fluorines to form hydrogen bonds that will affect molecular properties. Suprisingly, the evidence in the literature for hydrogen bonding to C-F groups is sparse. In a study from the Cambridge Database of 260 structures containing C-F bonds, there was only one convincing example of a C-F...H-O(N) hydrogen bond (Figure 2.10), along with two examples of possible shared hydrogen bonds.[35,36] It seems likely that a hydrogen bonding active hydrogen (typically, O-H and N-H) will prefer to hydrogen bond to less electronegative nitrogen or oxygen centers that possess more polarizable lone pairs. Since oxygen lone pairs, in particular, are in abundance in drug-macromolecule interactions, we can reasonably assume that hydrogen bonds to aromatic fluorines are unimportant. This does not rule out the possibility of other significant interactions involving C-F groups in biochemical environments, and coordination to alkali metals has been demonstrated and shown to be relevant to biological properties.[35]

While the hydrogen bonding properties of C-F groups themselves would seem to be largely unimportant, the strong electron withdrawing properties of fluorine may well affect the hydrogen bonding properties of other groups in the molecule and this may well affect drug potency.[37]

The dipole of the C-F bond itself might also play an important role in drug binding. An aromatic fluorine can successfully mimic a heterocyclic nitrogen in drug molecules such as quinolone antibiotics. A strong dipole in a drug molecule can be expected to seek out a complementary dipole at a receptor site so that situations in which the aromatic fluorine is actually intimately interacting with the biological macromolecule without actual hydrogen bonding can be envisaged.[38]

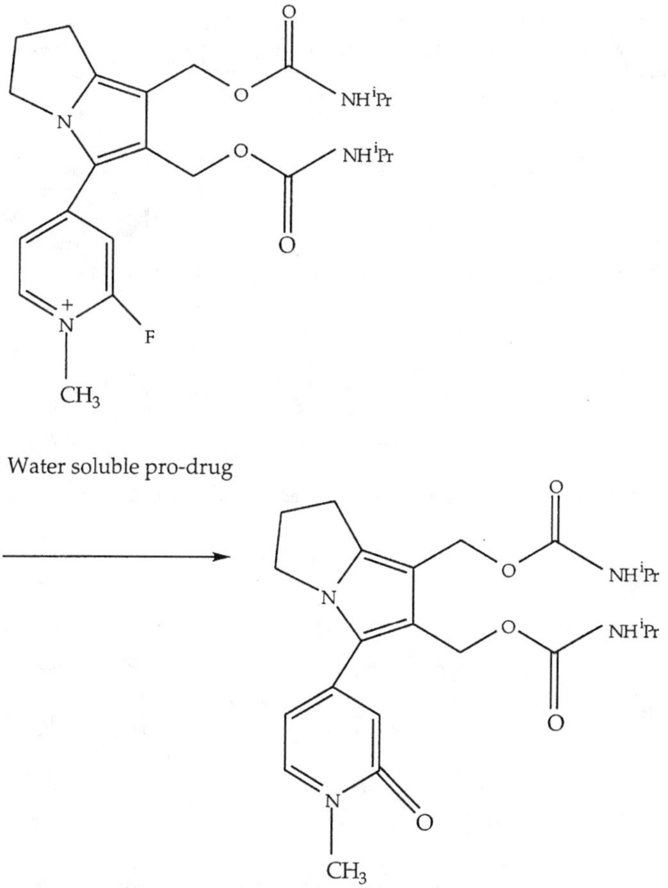

Water soluble pro-drug

FIGURE 2.11 Pharmacological application of fluorine as a leaving group.

2.4 CHEMICAL PROPERTIES OF FLUOROAROMATICS

2.4.1 Nucleophilic Substitution

The good leaving group ability of fluorine attached to electron-deficient aromatic systems can be utilized in pharmacological applications. A good example of this is where a pro-drug is used because of enhanced solubility, the pro-drug then being converted to the active form near the site of action (Figure 2.11).[39]

A more famous example of fluorine as a leaving group is from the highly electron-deficient 1-fluoro-2,4-dinitrobenzene (see Chapter 1), although even the fluorine in 1-fluoro-4-nitrobenzene is a relatively good leaving group (2.2×10^2 faster than 1-chloro-4-nitrobenzene in reaction with MeO⁻/MeOH). Aromatic fluorines are often sacrificed in multistep synthesis, such as in the synthesis of polyketones, including the high-performance polymer "PEEK" (Figure 2.12). It may also be more convenient to multiply fluorinate an aromatic system (thus avoiding the need for selectivity) and then selectively remove some of those fluorines (Figure 2.13).

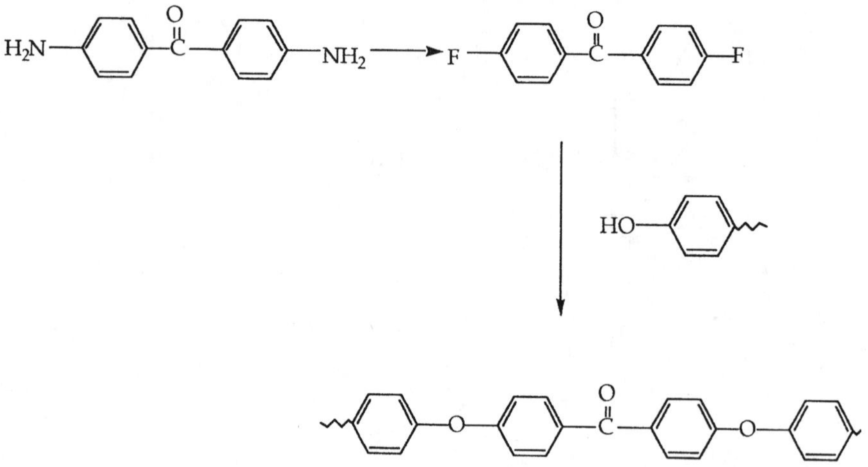

FIGURE 2.12   Use of fluoroaromatic intermediates in the synthesis of polyketones.

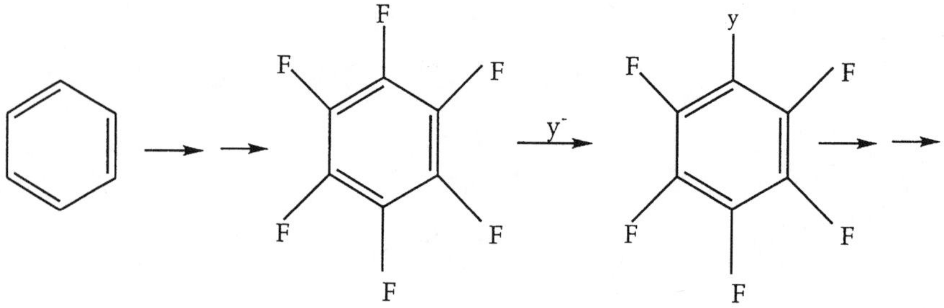

FIGURE 2.13   Example of sacrificial aromatic fluorination.

Fluorine is actually second only to the nitrogen in diazonium species as the best leaving group in nucleophilic aromatic substitutions preceded by the S_NAr mechanism.[40] The observed leaving group abilities in such systems ($N_2^+ > F > NO_2 > Cl$) is not the same as in aliphatic substitutions. The reason for the good leaving group ability of fluorine in S_NAr reactions is that the bond-forming first step in the reaction mechanism (the formation of the Meisenheimer intermediate) is the rate-determining step and this is enhanced by the high electronegativity of fluorine (the most electronegative of all the elements) inducing an especially high positive charge on the ring carbon to which it is attached (Figure 2.14). Nucleophilic substitution of fluorine by this route necessarily requires the presence of activating groups in the aromatic system. In reactions with MeO⁻ in MeOH, fluoro-2,4-dinitrobenzene is about 10^3 more reactive than 4-fluoronitrobenzene, but 4-fluoronitrobenzene is about 10^7 more reactive than fluorobenzene itself (i.e., the latter is effectively inert under such conditions).

The activation of fluoroaromatics for S_NAr reactions to proceed need not be mesomeric, so that hexafluorobenzene is only slightly less reactive than 4-fluoronitrobenzene. Indeed the nucleophilic displacement of fluorine from hexafluorobenzene has been widely used to make compounds of the general formula C_6F_5X (Figure 2.15).

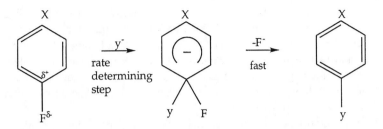

FIGURE 2.14   The S$_N$Ar mechanism in the nucleophilic substitution of fluorine.

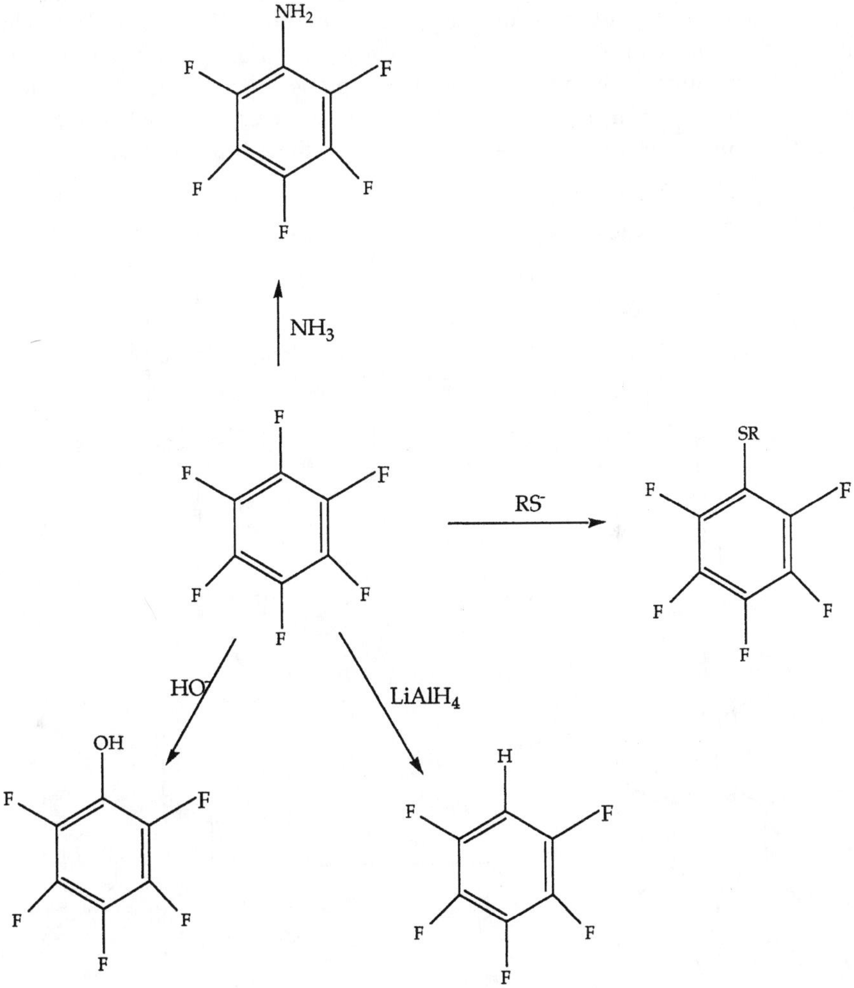

FIGURE 2.15   Reactions of hexafluorobenzenes.

Compounds of the formula C_6F_5X are also reactive towards nucleophilic substitution of fluorine and most X groups will direct methoxide attack for example, para to themselves (X = H, alkyl, CO_2Me, NO_2, NMe_2, etc.), although a few direct meta (X = NH_2 and OH) and some both meta and para (X = NHMe and OMe).[41] The effects of X on the reactivity of the aromatics are normal for nucleophilic aromatic substitutions,[42] with nitropentafluorobenzene being some 10^6 more reactive towards methoxide than pentafluorobenzene.

The observed orientations of attack and relative rates of reactions in nucleophilic substitution reactions of highly fluorinated aromatics can be largely explained by the Iπ effect (see Chapter 1) which will discourage attack of a nucleophile para to fluorine so that attack is directed para to X unless that substituent has an even greater Iπ effect (X = OH, NH_2). An alternative explanation is based on the activating effect of fluorine when it is ortho or meta to the point of attack. In the ortho position, it can activate by withdrawing electron density from the point of attack, whereas when it is meta it can stabilize the charge developing para to the point of attack in the Meisenheimer intermediate.[43] In reactions in aprotic media, attack may actually occur predominantly in the ortho position when the incoming nucleophile or its counter cation can be stabilized by association with the substituent X (Figure 2.16).

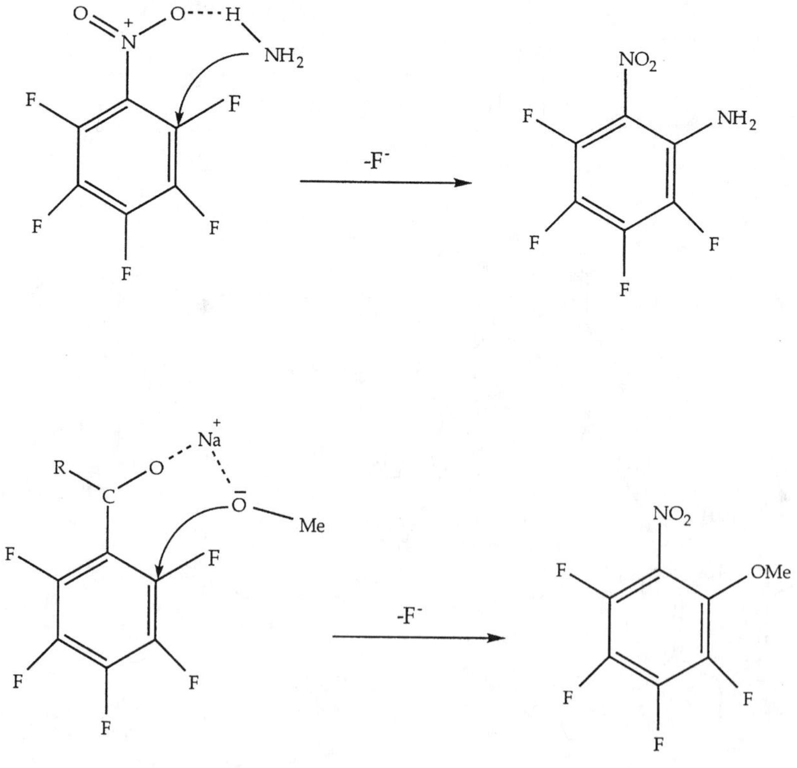

FIGURE 2.16 Substituent directed ortho substitution of aromatic fluorine.

Fluorine can also be displaced from aromatic systems by another route — the aryne mechanism. Here, however, fluorine compares unfavorably as a leaving group to the other halogens, as the rate-determining step is the breaking of the C-F bond, which is stronger than the other C-halogen bonds (Figure 2.17). The first step, the removal of the hydrogen ortho to the halogen, is actually fastest for fluorine, due to its relatively high inductive effect. The synthetic value of this reaction is rather limited, since the resistance of fluorine to act as a leaving group renders inactivated substrates such as fluorobenzene inert to attack by powerful bases such as sodium amide (which reacts with the other haloaromatics to give aniline). Heavily fluorinated metal aryls (e.g., tetrafluorophenyllithium[44]) can be used as useful intermediates in the synthesis of various fused ring systems.

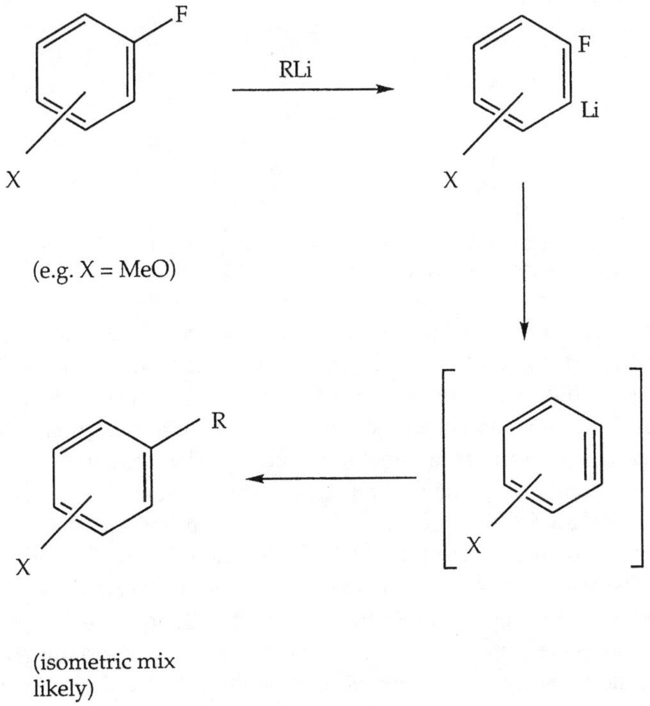

FIGURE 2.17   The aryne mechanism in the nucleophilic substitution of fluorine.

2.4.2 Electrophilic Substitution

Whereas the heavier halogens, as substituents always deactivate aromatics towards electrophilic attack, fluorine can activate or deactivate the aromatic nucleus. In all cases the effects are small compared to better-known activators (e.g., OMe) or deactivators (NO_2).[45] Generally the more reactive electrophiles, such as alkyl carbocations, R^+ in Friedel Crafts acylations, experience a deactivation of the aromatic substrate when fluorine is a substituent. This can be rationalized in terms of the transition state closely resembling the substrate and hence the electron density and nucleophilicity being reduced by fluorine. In the case of less reactive electrophiles such as Cl$^+$ or RCO$^+$, the intermediate will more closely resemble a Wheland intermediate and here fluorine can stabilize positive charge developing alpha to it via the Iπ effect (Figure 2.18).[46,47]

$$X-C_6H_5 \xrightarrow{E^+} X-C_6H_5-E$$

X =	H	E =	Et	kp* =	1.0
	F		Et		0.74
	Cl		Et		0.54
X =	H	E =	Cl	kp* =	1
	F		Cl		3.93
	Cl		Cl		0.31

FIGURE 2.18 Effect of fluorine on reactivity of aromatic substrates towards electrophilic attack. (*) represents the relative rates of attack of the electrophile at the para position.

It is important to note that at worst fluorine is only a mild deactivator and even heavily fluorinated aromatics are subject to electrophilic attack. Thus, polyfluorobenzenes up to pentafluorobenzene are subject to halogenation, Friedel Crafts alkylation and acylation, nitration, and other electrophilic substitution reactions.

Fluorine is a powerful para-director in electrophilic substitution. Indeed, of the common substituents, only *t*-butyl is more para-directing and that is presumably due to steric factors. Steric factors cannot be significant in the orientational effects of fluorine, which are again due to the preferential stabilization of positive charge by a para fluorine which means that intermediate I is more favored than intermediate II. Interestingly, the strong meta-deactivating effect of fluorine can cause unexpected orientational effects in substrates that possess competitive activating groups. Thus, 2-fluorotoluene is preferentially nitrated to give the 2-fluoro-5-nitrotoluene isomer (fluorobenzene is about 10^3 less reactive than benzene at the meta position, whereas toluene is slightly more reactive at the meta position than benzene).

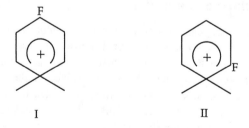

I   II

2.4.3 Radical Substitution

Radical reactions of fluorinated aromatics are of little synthetic value. Fluorobenzene reacts unselectively with phenyl radicals and some defluorination is also observed.[48,49] Phenyl radicals react more cleanly with hexafluorobenzene to give good yields of pentafluorobiphenyl[50,51] but perfluoroaryl radicals generally react to give complex product mixtures. Replacement of fluorine by radicals is quite commonly observed in such systems.[52]

2.4.4 Other Chemical Reactivity Effects

In the same way that fluorine substitution can have a long-range effect on the ability of a group to ionize or to interact with a biological receptor site, so it can have an effect on the chemical reactivity of distant groups. Thus, the reactivity of phenols, benzoic acids, etc. as acids is enhanced by ring fluorination, whereas the reactivity as anilines as bases for example, is reduced by ring fluorination.

2.5 ROUTES TO FLUOROAROMATICS

2.5.1 Halex Methods

The halex reaction in the context of fluorination is a nucleophilic aromatic substitution in which a chlorine, bromine, or (rarely) iodine atom X, commonly activated by an electron-withdrawing group Z in the ortho or para position, is displaced by fluorine upon reaction with a metal fluoride (e.g., NaF, KF) in a polar aprotic solvent (Figure 2.19). This traditional description is rather restrictive and it is possible through the use of more reactive fluorides or enhanced catalytic systems to extend the range of possible substrates to those not containing a resonance activating group, especially when the molecule possesses several inductively activating groups.

benzene)

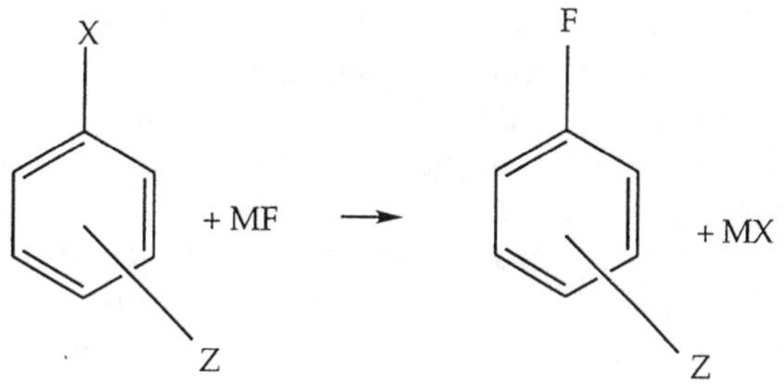

FIGURE 2.19 Traditional halex chemistry.

Halex chemistry is now established in manufacturing methods that go back more than 20 years. Some of the first commercial fluoroaromatics were based on 3-chloro-4-fluoroaniline which can be synthesized from 3,4-dichloronitrobenzene via the 3-chloro-4-nitrobenzene (Figure 2.20). The manufacture of this compound started in the UK in 1976 (Shell Chemicals UK, Stanlow).

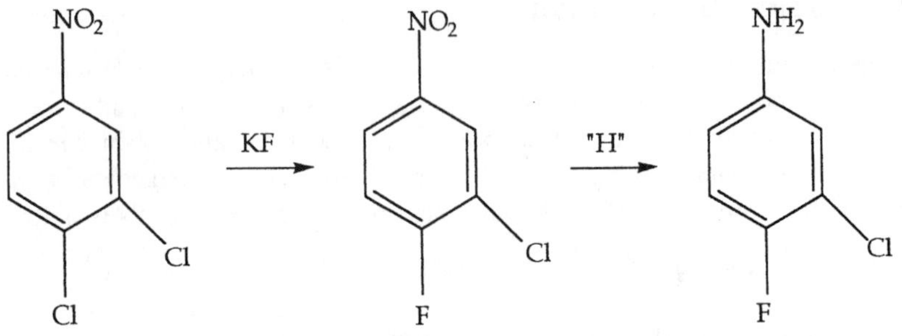

FIGURE 2.20   Manufacture of 3-chloro-4-fluoroaniline.

The basic halex technology employing KF as the fluoride source, a dipolar aprotic solvent such as dimethylacetamide or sulfolane, and a phase transfer catalyst (typically tetramethylammonium chloride)[53] has been successfully applied to the manufacture of a wide range of monofluoro and difluoro anilines, nitroaromatics, and benzonitriles (Figure 2.21).[54-56]

As mentioned above, the use of more reactive fluorides, such as cesium fluoride and quaternary ammonium fluorides, onium phase transfer catalysts, and crown ethers together with increasing reaction temperatures and, if necessary, pressure, widens the scope of reactive substrates to other than nitroaromatics and benzonitriles (Figure 2.22).[20,23,57-59] In some cases, competitive F$^-$ reactions such as decarbonylation of esters can result in reduced yields of the desired fluoroaromatic (Figure 2.23). In such cases, the undesired side reactions can be suppressed, at least partly, by the use of bulky alkyl groups to hinder attack of the F$^-$ at the carbonyl carbon. Generally, competitive reactions will be a greater problem with less electrophilic aromatics. Thus, whereas even sterically hindered monochlorobenzoates give low yields of fluoroaromatic products under halex conditions, substrates with greater numbers of chlorine substituents (or other electron-withdrawing groups) give better results.[57]

The drive towards cleaner chemical technologies and the reduction of waste at source will put increasing pressure on manufacturers to carefully review all aspects of their processes. In this context solvents will be given special consideration and replacements will be sought for dipolar aprotic solvents, which are toxic, prone to thermal and chemical decomposition, and difficult to recover and recycle efficiently. Because of this, increasing attention will be given to developing effective halex systems that employ hydrocarbon or other environmentally acceptable solvents or indeed those that require no solvent at all. The use of phase transfer catalysts is likely to be especially important in these developments. In a typical halex reaction (Figure 2.24) the rate acceleration achieved using the catalyst Ph$_4$PBr is greater when the reaction is carried out in less polar solvents. The effect of this catalyst on the halex reaction of 4-chloro-3-nitrobenzotrifluoride is to increase the rate of reaction by a factor of 4.4 in the solvent sulfolane, but by a factor of 200 in the less polar acetonitrile, a solvent that is generally less efficient in halex reactions. Individual catalytic rate enhancements and their sensitivity to solvent are less with more reactive

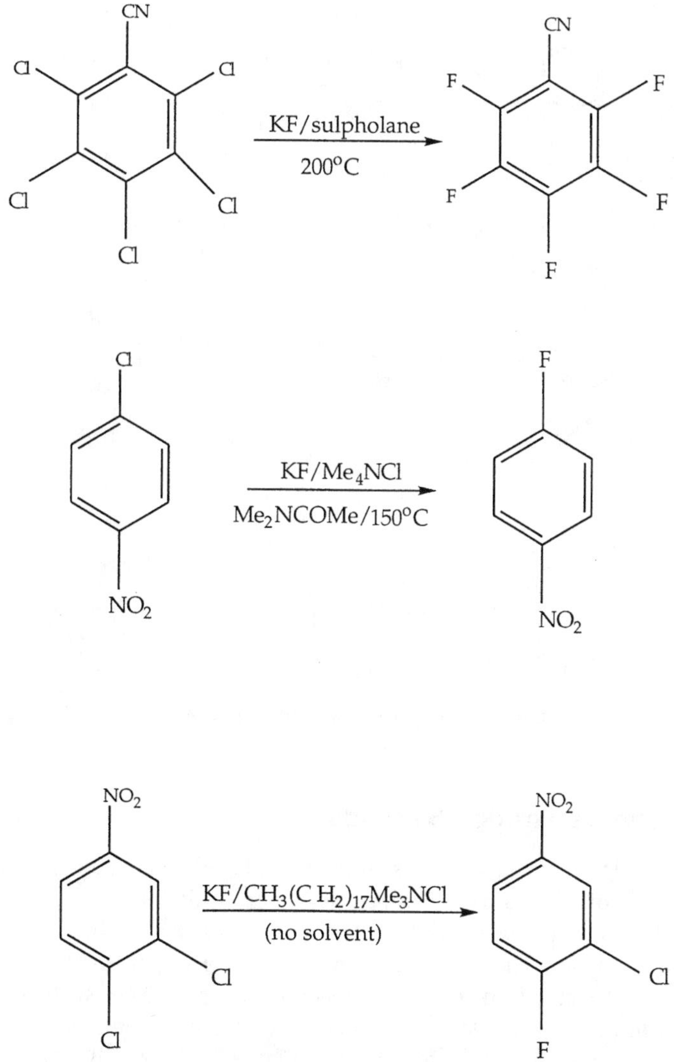

FIGURE 2.21 Examples of halex reactions using traditional fluoride systems.

substrates.[60] This is probably because a major role of the dipolar aprotic solvent in these reactions is the stabilization of the anionic (Meisenheimer) intermediates, and in the absence of such solvents, onium cations may be able to perform this role.

The more soluble and reactive onium fluorides (tetrabutylammonium fluoride, tetramethylammonium fluoride, tetraphenylphosphonium bifluoride, and others) enable many relatively difficult halex reactions to be carried out under mild conditions, although the high basicity of these reagents may cause complications through competitive reactions and solvent decomposition, as well as decomposition (e.g., via Hoffman degradation) of the reagent itself.[26,29,31,61]

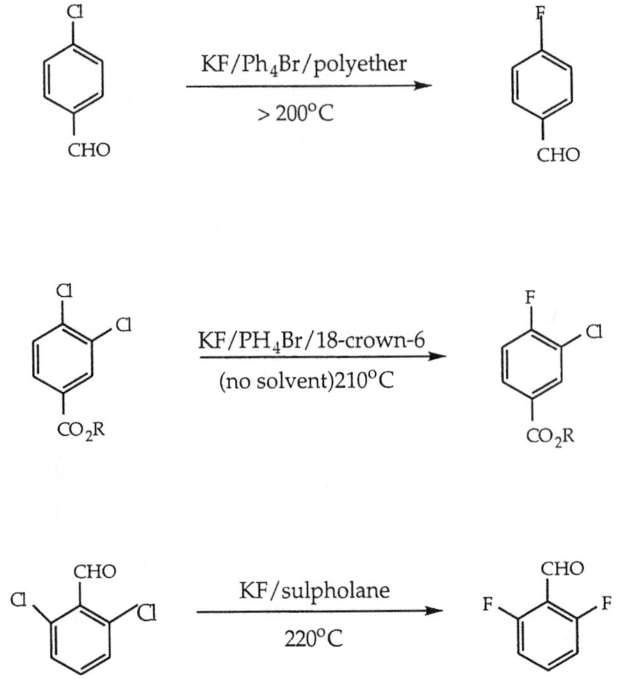

FIGURE 2.22   Some more difficult halex reactions.

2.5.2 Fluorodenitration Methods

From being originally considered as an undesirable side reaction that could lead to reduced yields of the desired halex product, fluorodenitration is rapidly becoming established as a viable alternative route to halex for the synthesis of selectively fluorinated aromatics. There are a growing number of reports of fluorodenitration, both in terms of methods to achieve the reaction and in terms of synthesizing target molecules. In the latter context, it has been used for the synthesis of useful pharmaceutical and agrochemical intermediates (Figure 2.25).[21,62] The starting material for the synthesis of such a useful intermediate, 3-chloro-4-fluoronitrobenzene, can be prepared by oxidation of the readily available 2-chloro-4-nitroaniline with hydrogen peroxide.[62]

The reactive substrate, 4-chloro-3,5-dinitrobenzotrifluoride, readily undergoes halex and fluorodenitration to give the difluoro product which can be used en route to aryl pyridone and pyrimidinone with useful insecticide properties.

The expansion in the field of ^{18}F-labeled radiopharmaceuticals requires the development of fast and efficient methods of synthesis of organofluorine compounds using fluorides (of the available fluorinating agents, only [18]F$^-$ can be conveniently prepared in high yields without the addition of carriers). Fluorodenitration, which is generally faster than halex under conditions of comparable substrate activation, can be effective in this context and numerous rapid syntheses of ^{18}F-labeled aromatics

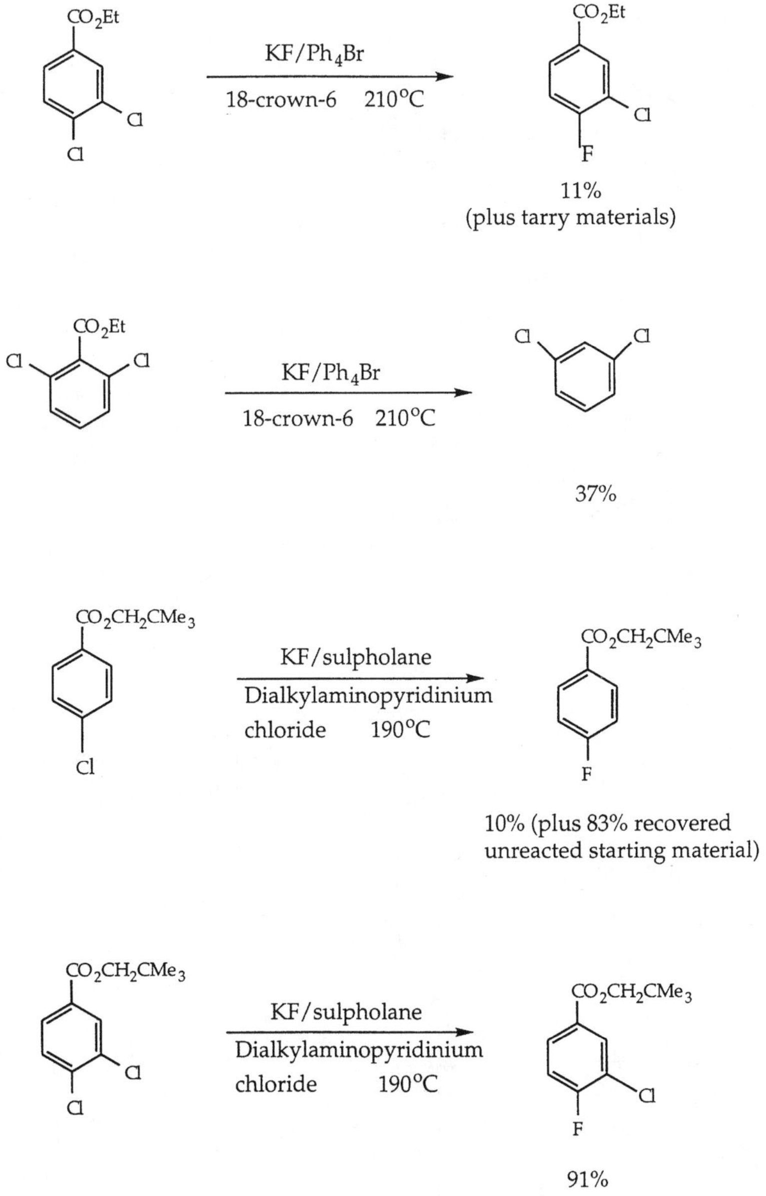

FIGURE 2.23   Competitive halex and decarbonylation.

have been reported (Figure 2.26).[61,63] The use of microwaves to give rapid intense heating can be used to further reduce reaction times and hence increase radiochemical yield.[63]

The better leaving group ability of nitro compared to chlorine means that intrinsically less activated substrates may be viable starting materials for fluoroaromatics via fluorodenitration. The reactivity of nitrophthalic anhydrides, for example,

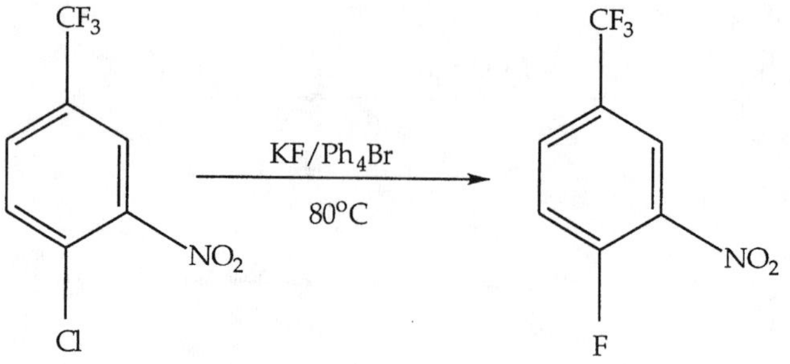

FIGURE 2.24   Phase transfer catalyzed halex.

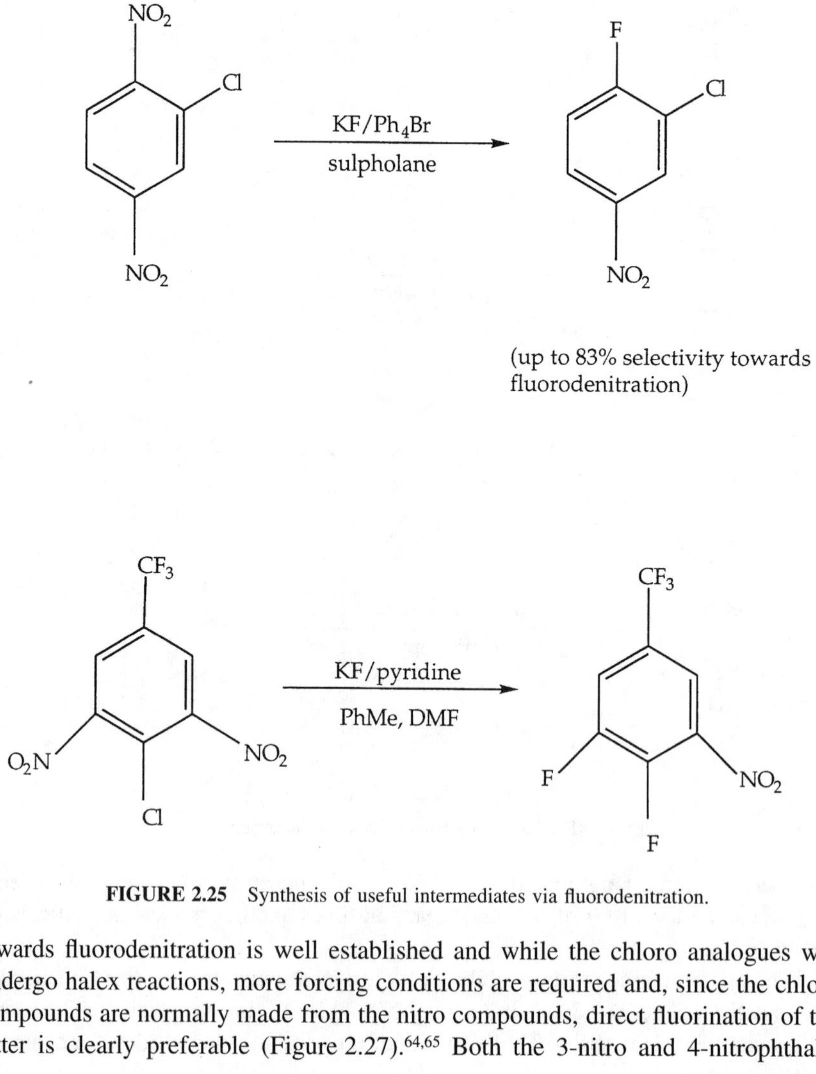

(up to 83% selectivity towards
fluorodenitration)

FIGURE 2.25   Synthesis of useful intermediates via fluorodenitration.

towards fluorodenitration is well established and while the chloro analogues will
undergo halex reactions, more forcing conditions are required and, since the chloro
compounds are normally made from the nitro compounds, direct fluorination of the
latter is clearly preferable (Figure 2.27).[64,65] Both the 3-nitro and 4-nitrophthalic

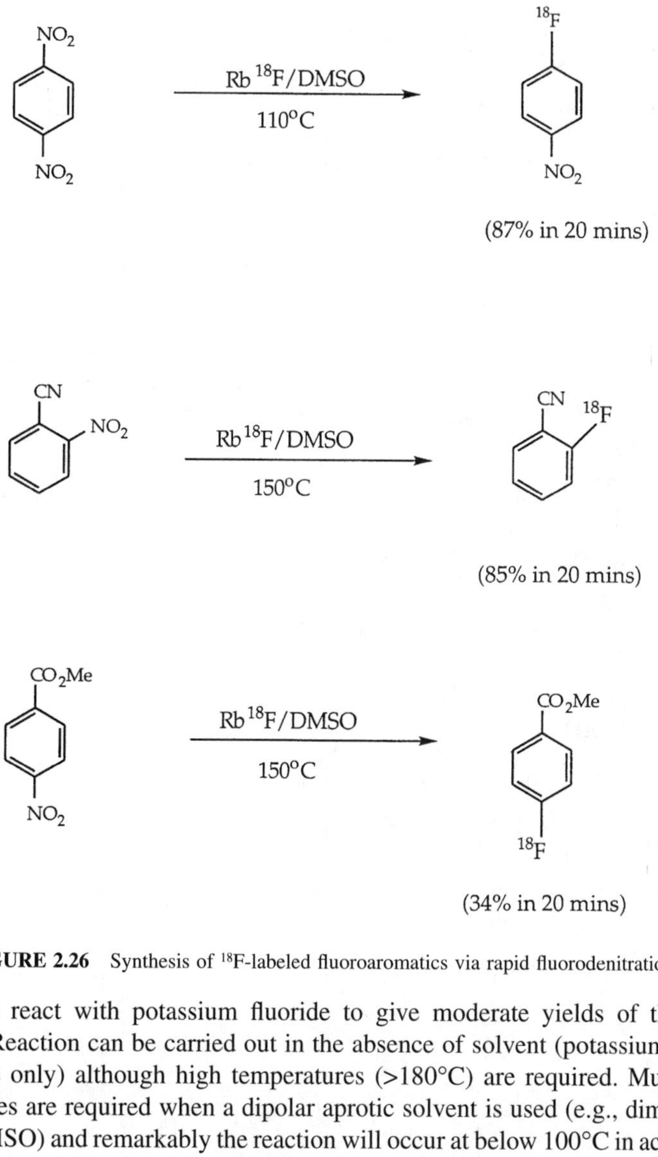

FIGURE 2.26 Synthesis of ^{18}F-labeled fluoroaromatics via rapid fluorodenitration.

anhydrides react with potassium fluoride to give moderate yields of the fluoro products. Reaction can be carried out in the absence of solvent (potassium fluoride + substrate only) although high temperatures (>180°C) are required. Much lower temperatures are required when a dipolar aprotic solvent is used (e.g., dimethylsulfoxide, DMSO) and remarkably the reaction will occur at below 100°C in acetonitrile in the presence of the cation solvating 18-crown-6 (one of the more impressive examples of a crown-assisted fluoride reaction). Interestingly, substituted phthalic acids (potassium salts of) can also be formed in these reactions and this is thought to be due to reaction of the eliminated nitrite ion with the anhydride (Figure 2.28).[65] This reaction is reminiscent of one of the more popular modern methods of trapping nitrite (see below).

Activation by two carbonyls as in the anhydrides above is important but not essential for facile fluorodenitration. Thus, nitro-substituted indanediones readily undergo fluorodenitration under mild conditions (KF/DMSO, 120°C), whereas the indanones are unreactive under such conditions (Figure 2.29).[66]

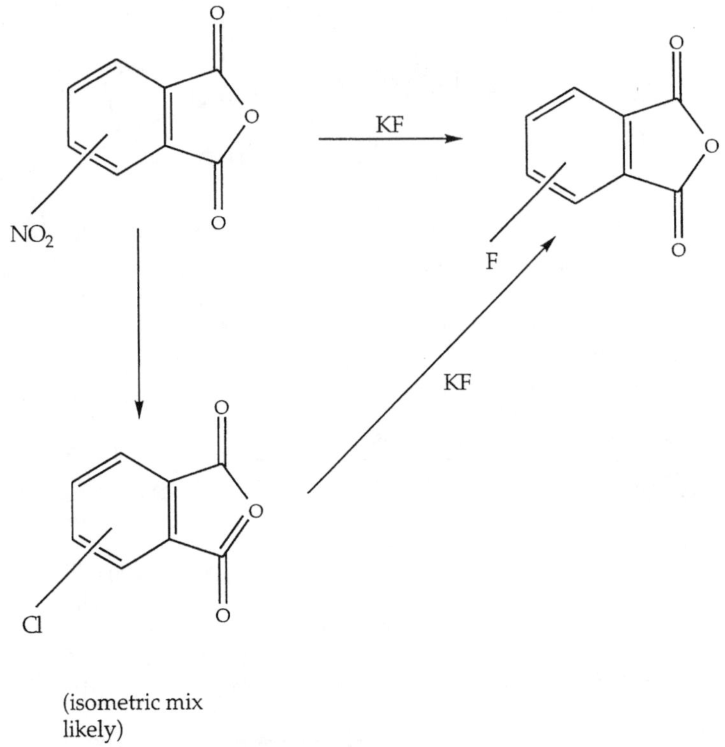

FIGURE 2.27   Routes to fluorophthalic anhydrides.

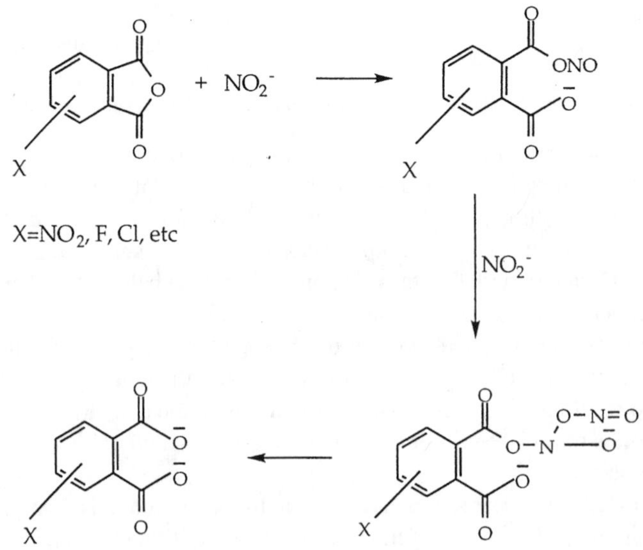

FIGURE 2.28   Side reaction of phthalic anhydrides with nitrite ion eliminated in denitration reactions.

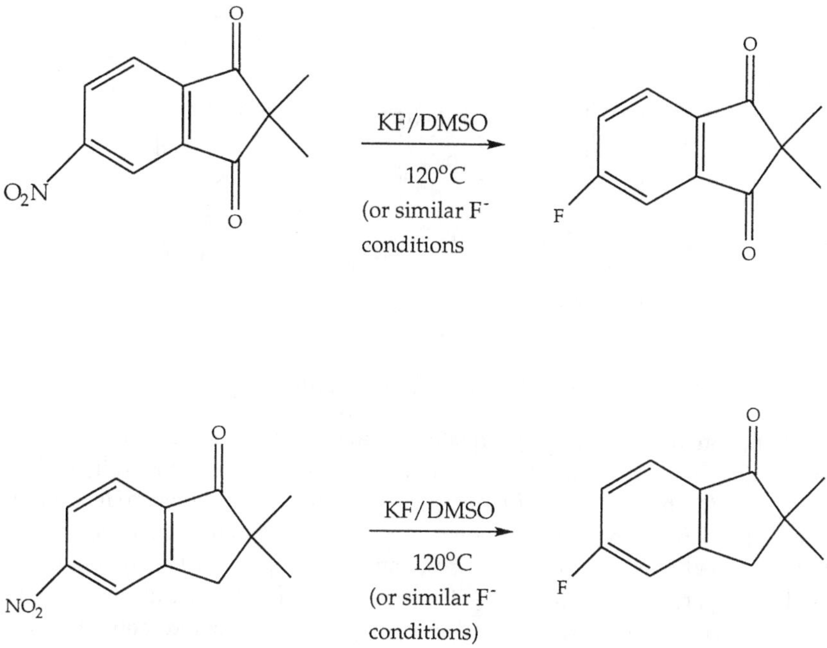

FIGURE 2.29   Reactions of nitroindanediones and nitroindaneones with fluoride.

Activation by a single carbonyl group may enable effective fluorodenitration. Thus, 2-nitrobenzaldehyde will react with more potent fluorodenitration reagents such as tetramethylammonium fluoride ("TMAF") to give reasonable yields of the fluorobenzaldehyde product (Figure 2.30).[27] Nitro-aromatic esters can also undergo fluorodenitration so that 4-nitro-methylbenzoate reacts with rubidium fluoride for example, although forcing conditions are required and yields may be poor (Figure 2.31).[61] Fluorodenitration of benzoate esters occurs much more readily when further activation is provided by a mild inductive activator such as chlorine.[67] Here of course, there is now the possibility of halex entering into competition with fluorodenitration.

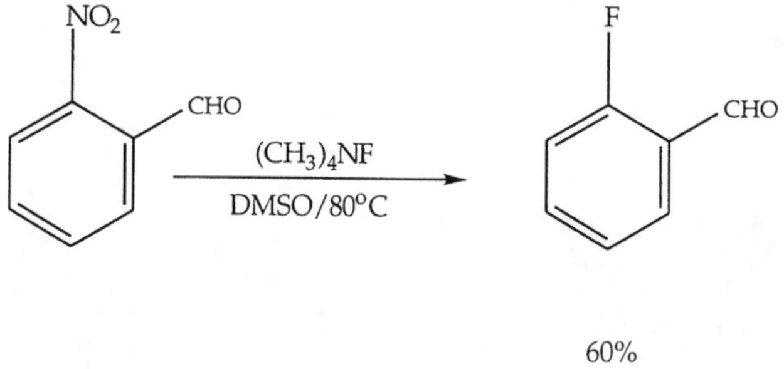

FIGURE 2.30   Fluorodenitration of 2-nitrobenzaldehyde.

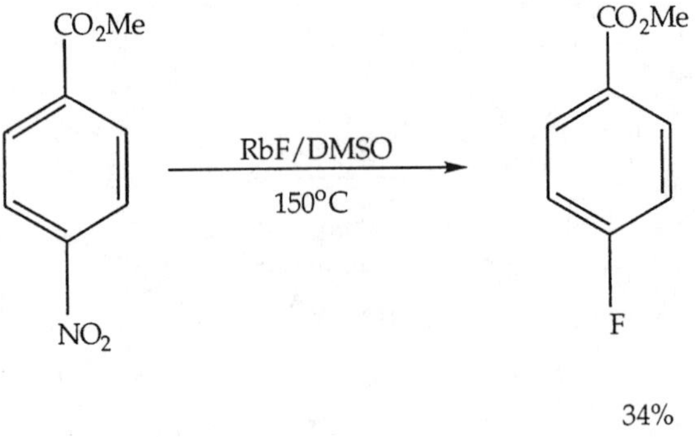

FIGURE 2.31 Fluorodenitration of a benzoate ester.

Studies on the reactivity of the model substrate 4-chloro-2-nitro-methylbenzoate (Figure 2.32) with different KF reaction systems showed that the KF/Ph$_4$Br/DMSO system gave the best yields of fluoroaromatic product. The major product was that resulting from fluorodenitration accompanied by trace amounts of the halex product and the product resulting from decarboxylations (this product becomes significant at higher temperatures in sulfolane — Figure 2.32). Fluorodenitration also dominates the reactions of many other activated aromatics containing chloro and nitro groups (Figure 2.33). The nucleophilic fluorination chemistry of the corresponding trifluoro-methylsulfones is more complex (Figure 2.34). 4-Chloro-2-nitro-trifluoromethyl-phenylsulfone reacts to give predominantly the fluorodenitration product but with significant amounts of the difluorosulfone and the fluorodesulfonylation product. The isomer, 2-chloro-4-nitro-trifluoromethylphenylsulfone, also gives mostly the product from fluorodenitration, although the relative efficiency of the process is dependent on the reaction conditions (Table 2.5).[67] Interestingly, excellent selectivity to the fluorodenitration product can be achieved either at very high temperatures or by using the highly reactive tetrabutylammonium fluoride, at very low temperatures.

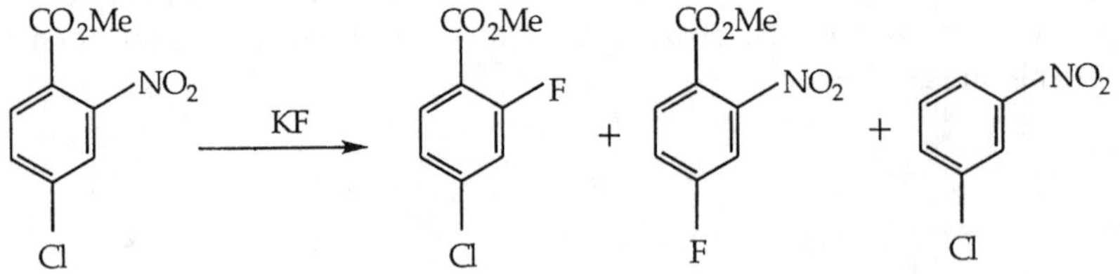

Solvent/temperature (°C)			
Sulpholane/220	27	—	66
Sulpholane/180	45	1.3	38
DMAc/130	31	5	2
DMSO/130	72	1	3

FIGURE 2.32 Reactions of 4-chloro-2-nitro-methylbenzoate with fluoride.

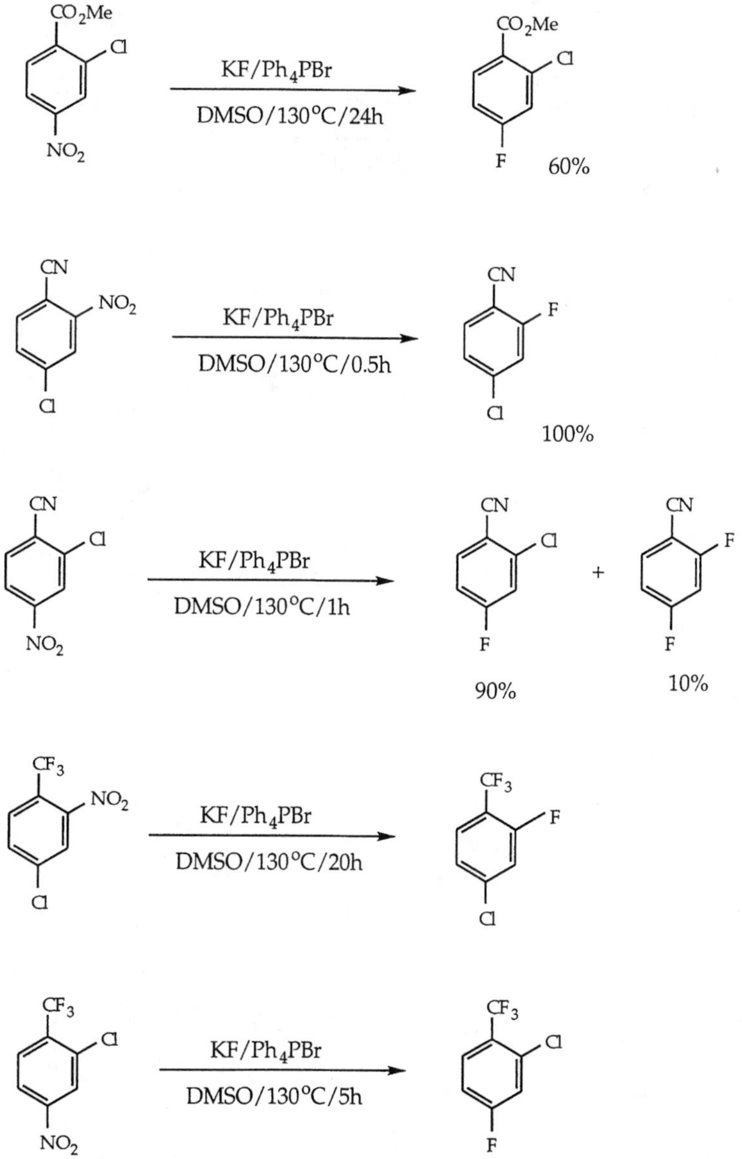

FIGURE 2.33 Reactions of activated chloronitroaromatics with fluoride.

In many cases, fluorodenitration results in complex product mixtures and rather poor isolated yields of the desired fluoroaromatic product. Side reactions leading to the formation of phenolic and diaryl ether by-products are particularly common and may not be easily recognized by gas chromatographic analysis, resulting in misleading apparent reaction efficiencies which are only clarified on separation and isolation of the fluoroaromatic. The by-products are believed to be due to reactions of the displaced nitrite ion — either by direct attack by the anion on the aromatic or by

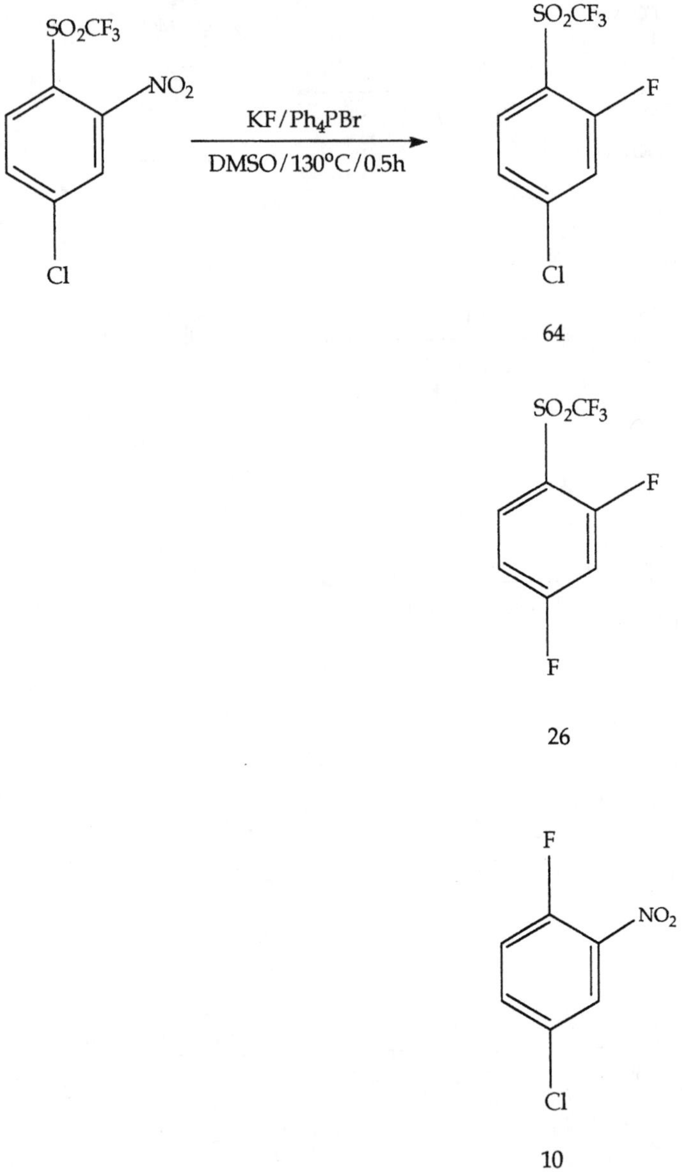

FIGURE 2.34 Reactions of 4-chloro-2-nitro-trifluoromethylphenylsulfone with fluoride.

initial decomposition of the nitrite to an oxide which can then attack the aromatic (Figure 2.35). The exact fate of the nitrogen is unknown, although brown fumes of nitrogen oxides are often observed in fluorodenitrations (although curiously there are many effective fluorodenitrations where brown fumes are not observed).[68] To help overcome this problem, phthaloyl dichloride (PDC) can be used as a nitrite ion trap (Figure 2.36).[69,70] This is an effective methodology enabling good yields of fluoroaromatics to be obtained via fluorodenitration (Figure 2.37).

Table 2.5 Fluorination of 2-chloro-4-nitro-trifluoromethylphenylsulfone Under Different Conditions

Fluoride	Solvent/temp	Products (%)		
		Products (%)		
		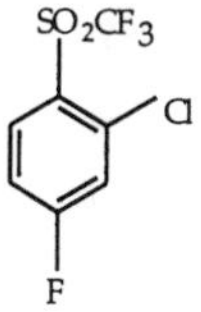	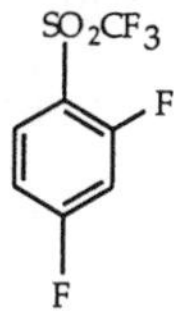	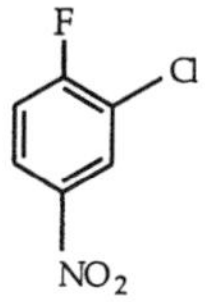
Reagent	(°C)/time(min)			
TBAF	THF/-78/25	98	0	2
KF	Sulfolane/230/1	100	0	0
TBAF	THF/20/5	67	9	24
KF	Sulfolane/170/5	80	10	10
KF	DMSO/130/5	78	5	17

$$ArF + NO_2^- \longrightarrow ArONO \longrightarrow ArO^- + NO$$

$$2KNO_2 \longrightarrow K_2O + N_2O_3$$

$$ArF + K_2O \longrightarrow ArO^-$$

FIGURE 2.35 Side reactions due to the nitrite ion displaced in fluorodenitration reactions.

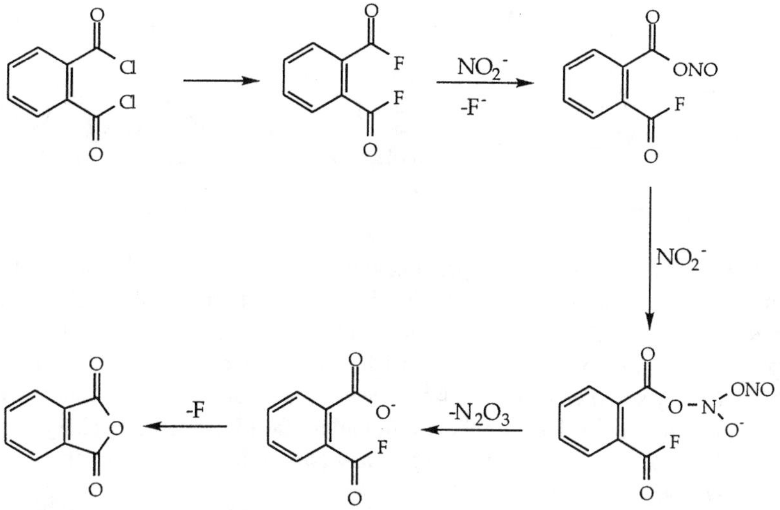

FIGURE 2.36 Possible mechanism for the reaction of phthaloyl difluoride (from the dichloride) with nitrite.

A particularly valuable aspect of utilizing a nitrite ion trap to capture displaced nitrite ion is that more difficult fluorodenitrations such as those of meta-substituted substrates become viable. It is often the case that slower reactions result in lower product yields due to side reactions of the nitrite ion.

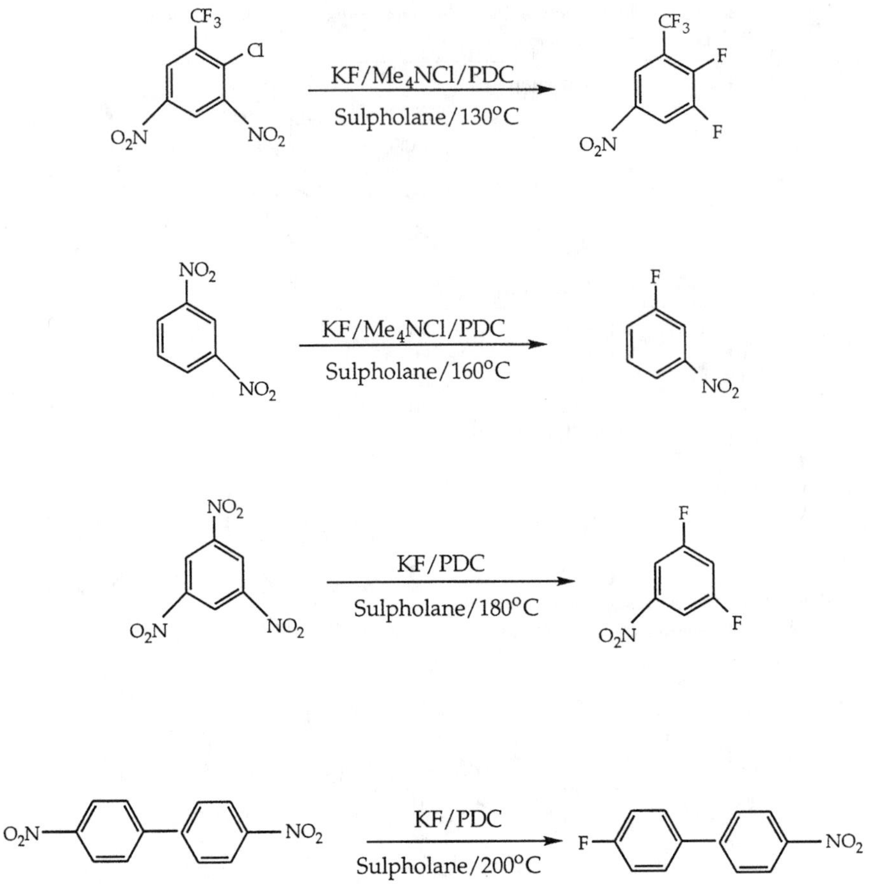

FIGURE 2.37 Fluorodenitration in the presence of phthaloyl dichloride.

An alternative method of improving the yields of fluorodenitration is to use tetramethylammonium fluoride as the fluoride reagent (Figure 2.38).[27] The cation is apparently able to stabilize the nitrite ion and reduce its reactivity through strong ion pairing. Tetramethylammonium fluoride is more reactive than the alkali metal fluorides (being soluble in polar solvents), but less reactive than the heavier tetraalkylammonium fluorides. Tetrabutylammonium fluoride is an especially reactive source of fluoride and rapidly reacts with many nitroaromatics,[31] although yields are often rather poor.

Fluorodenitration can now be considered as an alternative route to halex for nucleophilic fluorination. It allows access to more difficult products, including metafluoroaromatics, and nitro groups will normally be substituted in preference to chlorine under conditions of comparable activation. The greatest drawback of fluorodenitration is the risk of by-products and, for other than very quick reactions, an additional reagent that can act as a nitrite ion trap should be present (although this is not always apparently effective) or tetramethylammonium fluoride should be used to act as a simultaneous source of fluoride and nitrite trap.

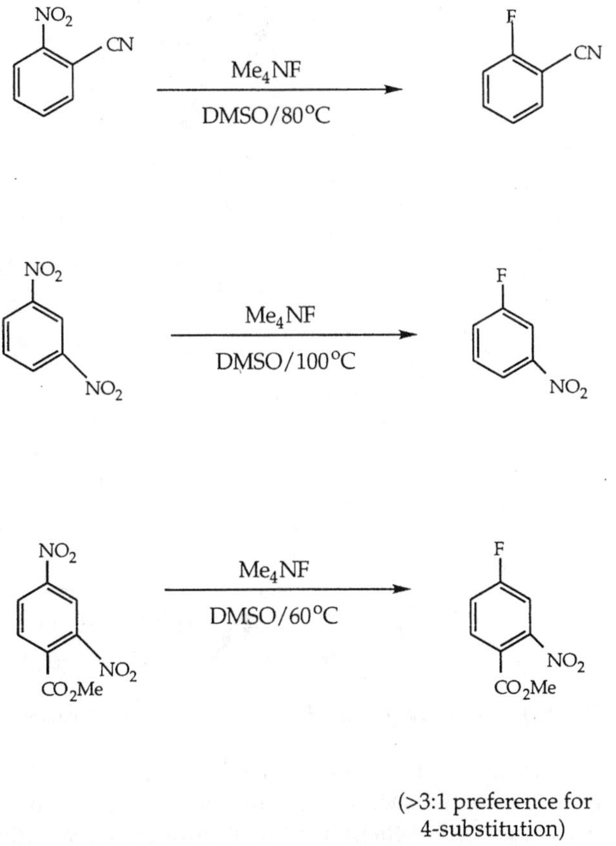

(>3:1 preference for
4-substitution)

FIGURE 2.38 Fluorodenitration using tetramethylammonium fluoride.

2.5.3 Other Aromatic Nucleophilic Fluorinations

Several aromatic substituents other than chloro and nitro have been shown to be susceptible to nucleophilic substitution by fluoride. Alternative leaving groups include phenylthio (ArSPh),[71] diaryliodonium (Ar$_2$I$^+$ X$^-$),[72] dimethylsulfonium (ArSMe$_2^+$X$^-$), sulphonyl (ArSO$_2$Ar), as well as the heavier halogens (haloaromatics). While fluorination of some of these aryl substrates using fluoride may be facile, their high cost, limited availability, or difficulty of synthesis precludes their general use for the synthesis of fluoroaromatics.

One other F$^-$-based aromatic fluorination worth including here is the Bamberger rearrangement.[73,74] In this reaction the partial reduction products of nitroaromatics, N-arylhydroxylamines, react with anhydrous hydrogen fluoride to produce the corresponding 4-fluoroanilines.[75-78] This is potentially a very interesting method for converting unactivated nitroaromatics to the valuable fluoroanilines, but in practice, competitive reduction of the N-arylhydroxylamine to the corresponding unfluorinated aniline can lead to much reduced yields in one-pot reduction/fluorination

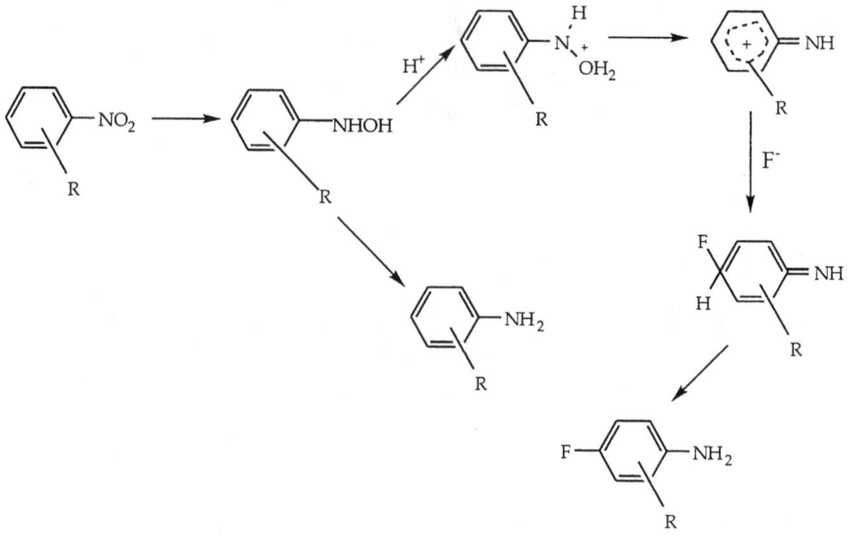

(R = 2-, or 3-, CH$_3$, Cl, C$_2$H$_5$CO$_2^-$, etc;
product yield typically 30 - 60%)

FIGURE 2.39   The Bamberger rearrangement for synthesizing 4-fluoroanilines.

reactions (Figure 2.39). Even reactions starting from the *N*-arylhydroxylamine and HF commonly give poor product yields along with tarry, unworkable material accounting for the balance.[78] 4-Substituted products do not give a fluoro product. 2-Fluoroanilines are not observed.

2.6 SYNTHETIC METHODS

Some of the most widely used sources of fluoride along with comments on their properties were given earlier (Table 2.2). It is generally true to say that the more reactive the fluoride, the greater the associated handling difficulties. The more reactive metal fluorides (CsF, RbF) are very hygroscopic, making handling difficult and requiring extensive preactivation. The highly reactive onium fluorides (especially tetrabutylammonium fluoride, TBAF) are effectively impossible to dry (tetramethylammonium fluoride, TMAF, is the most reactive quaternary ammonium fluoride that can be made anhydrous, although only after considerable effort) and are thermally unstable (with TBAF decomposing at > approximately 50°C).

It is sometimes believed that a trace of water can be beneficial to fluoride activity, presumably due to the lowering of the lattice energy through hydration at the crystal-liquid interface. This may commonly be the case in reactions employing fluorides as bases (where any hydroxide generated can also act as a base) but is a more risky premise in nucleophilic fluorinations where any hydroxide can act as a competitive nucleophile, resulting in the formation of stable phenols/ethers and a reduction in the yield of the desired fluoroaromatic product.

Phase transfer catalysts are commonly used in reactions employing KF as the fluoride source. A phase transfer catalyst mole ratio (with respect to KF) of about 5% is reasonable, although it may be necessary to decrease this to make the process viable, since the catalyst can rarely be efficiently recovered. An exception to this is tetraphenylphosphonium bromide, which is extremely thermally stable (at least up to temperatures of about 300°C) and can be left in the residue after distillation of the products and solvent, and then recovered by dissolution in a moderately polar solvent such as tetrahydrofuran or hot toluene. The only other thermally stable phase transfer catalyst in common use is tetramethylammonium chloride (TMACl) and, while this is considerably cheaper, it is also a poorer catalyst.

Nucleophilic aromatic fluorinations have traditionally been carried out in dipolar aprotic solvents and these will be difficult to replace, although no-solvent reactions will become more common. These solvents are themselves subject to attack by the more reactive fluorides. Dimethylsulfoxide, dimethylformamide, and many other amide solvents are especially prone to attack and should not be used with fluorides such as TBAF. Sulfolane is a little more stable but can still be decomposed by F^-. Side products resulting from solvent decomposition can include the aromatic hydro-defluorination products. Generally it is advisable to use the more reactive (highly soluble) fluorides with less reactive solvents — tetrahydrofuran and acetonitrile can be used over short periods. In all these cases it is generally advisable to dry the solvents before use and to be aware of the possible formation of side products resulting from solvent participation/breakdown.

The more common KF-based reactions are heterogeneous and based on a fairly low surface area reagent. These reactions therefore require thorough mixing, which can also help to keep the surface of unreacted KF clean. Typically reactions also require heating and microwave activation can be used to reduce reaction times.[63] This is especially important in the synthesis of ^{18}F-labeled aromatics as radiopharmaceuticals, but might also be useful in fluorodenitration reactions, where longer reaction periods generally result in lower isolated product yields due to the build up of ether products. It can also be practical in heated reactions to arrange for the continuous removal of the more volatile fluoroaromatic product (the fluoroaromatic is always the most volatile aromatic). This method has the advantages of combining separation with reaction and of reducing the possibility of side reactions destroying the valuable fluoroaromatic product. The use of nitrite ion traps (e.g., phthaloyl dichloride) may also help to reduce side-product formation in fluorodenitrations, although they are not consistently effective.[67] Whenever possible, short reaction periods should be used in fluorodenitrations.

It is also important to keep in mind the potential hazards of working with fluorides and hydrogen fluoride in particular. Extreme care should be exercised when working with HF and all reactions should be carried out in efficient fume hoods. Most other fluoride reagents are less hazardous but they should be regarded as being very toxic and all fluoride reaction systems should be assumed to contain some HF.

REFERENCES

1. F. Swarts, *Bull. Acad. R. Belg.,* 1898, 35, 301.
2. H.B. Gottlieb, *J. Am. Chem. Soc.,* 1936, 58, 532.
3. G.C. Finger and C.W. Kruse, *J. Am. Chem. Soc.,* 1956, 78, 6034.
4. W. Prescott, *Chem. Ind.,* 1978, 56.
5. J.H. Clark, M.J. Robertson, D.K. Smith, A. Cook and C. Streich, *J. Fluorine Chem.,* 1985, 28, 161.
6. R.D. Chambers, *Fluorine in Chemistry,* Wiley, New York, 1973.
7. K.F. Bunnett and R.J. Morath, *J. Am. Chem. Soc.,* 1955, 77, 5051.
8. L.D. Starr and G.C. Finger, *Chem. Ind.,* 1962, 1328.
9. G.C. Finger and L.D. Starr, *J. Am. Chem. Soc.,* 1959, 81, 2674.
10. G.C. Finger, L.D. Starr, D.R. Dickerson, H.S. Gutowsky and J. Hamer, *J. Org. Chem.,* 1963, 28, 1666.
11. F. Mutterer and C.D. Weis, *Helv. Chim. Acta,* 1976, 59, 229.
12. H.C. Fielding, L.P. Gallimore, H.L. Roberts and B. Title, *J. Chem. Soc. (C),* 1966, 2142.
13. J. Gisley, *Chem. Soc. Rev.,* 1980, 91, 9.
14. J.H. Clark, *Chem. Rev.,* 1980, 80, 429.
15. N. Ishikawa, T. Kitazume, T. Yamazaki, T. Mochida and T. Tatsuno, *Chem. Lett.,* 1981, 761.
16. J.H. Clark, A.P. Kybett and D.J. Macquarrie, *Supported Reagents: Preparation, Analysis and Applications,* VCH, New York, 1992.
17. T. Ando, S.J. Brown, J.H. Clark, D.G. Cork, T. Hanafusa, J. Ichihara, J.M. Miller and M.S. Robertson, *J. Chem. Soc. Perkin Trans. 2,* 1986, 1133.
18. J.H. Clark, A.J. Hyde and D.K. Smith, *J. Chem. Soc. Chem. Commun.,* 1986, 791.
19. J. Ichihara, T. Matsuo, T. Hanafusa and T. Ando, *J. Chem. Soc. Chem. Commun.,* 1986, 793.
20. J. Yoshida and J. Kimura, *J. Fluorine Chem.,* 1989, 44, 291.
21. J.H. Clark and D.J. Macquarrie, *Tetrahedron Lett.,* 1987, 28, 111.
22. J.H. Clark and J. Miller, *J. Chem. Soc. Chem. Commun.,* 1982, 1318.
23. J. Joshida and J. Kimura, *Chem. Lett.,* 1988, 1355.
24. E. Kysela and R. Braden, Gen. Offen. DE 3, 827, 436 (*Chem. Abstr,* 1990, 113, 40127e).
25. J.A. Wilkinson, *Chem. Rev.,* 1992, 92, 505.
26. D.P. Cox, J. Terpinski and W. Lawrynowqicz, *J. Org. Chem.,* 1984, 49, 3216.
27. N. Boechat and J.H. Clark, *J. Chem. Soc. Chem. Commun.,* 1993, 921.
28. P. Bosch, F. Camps, E. Chamorro, V. Gasol and A. Guerrero, *Tetrahedron Lett.,* 1987, 28, 4733.
29. S.J. Brown and J.H. Clark, *J. Fluorine Chem.,* 1985, 251.
30. S.J. Brown and J.H. Clark, *J. Chem. Soc. Chem. Commun.,* 1985, 672.
31. J.H. Clark and D.K. Smith, *Tetrahedron Lett.,* 1985, 26, 2233.
32. K.O. Christie, W.W. Wilson, R.D. Wilson, R. Bau and J. Feng, *J. Am. Chem. Soc.,* 1990, 112, 7619.
33. J.T. Welch, *Tetrahedron,* 1987, 43, 3123.
34. D.T.W. Chu and P.B. Fernandes, *Antimicrob. Agents Chemother.,* 1989, 33, 131.
35. P. Murray-Rust, W. C. Stallings, C.T. Monti, R.K. Preston and J. P. Glusker, *J. Am. Chem. Soc.,* 1983, 105, 3206.
36. S. Doddrell, G. Wenkert and P.V. Demorro, *J. Mol. Spectrosc.,* 1969, 52, 162.
37. U. Madren, B. Ebert, P. Krogsgaard-Larsen and E.H.F. Wong, *Eur. J. Med.,* 1992, 27, 479.
38. W. L. Kirk and C.R. Creveling, *Med. Res. Rev.,* 1984, 4, 189.
39. W.K. Anderson, D.C. Dean and T. Endo, *J. Med. Chem.,* 1990, 33, 1667.
40. J.F. Bunnett and R.E. Zahler, *Chem. Rev.,* 1951, 49, 273.
41. L.S. Kobrina, *Fluorine Chem. Rev.,* 1974, I, 1.
42. J. Miller, *Aromatic Nucleophilic Substitution,* Elsevier, Amsterdam, 1968.
43. R.D. Chambers, M.J. Seabury, D.L.H. Williams and N. Hughes, *J. Chem. Soc. Perkin Trans. 1,* 1988, 251.
44. D.J. Burton, Z.Y. Yang and P.A. Marten, *Tetrahedron,* 1994, 50, 2993.
45. R.O.C. Norman and R. Taylor, *Electrophilic Substitution in Benzenoid Compounds,* Elsevier, Amsterdam, 1965.
46. G.A. Olah, Y.K. Mo and J. Halpern, *J. Am. Chem. Soc.,* 1972, 94, 2034.
47. L.M. Stock and H.C. Brown, *Adv. Phys. Org. Chem.,* 1963, 1, 35.

48. P. Lewis and G.H. Williams, *J. Chem. Soc. (B),* 1969, 120.
49. J.F.B. Sandall, R. Bolton and G.H. Williams, *J. Fluorine Chem.,* 1973, 3, 35.
50. P.A. Claret, G.H. Williams and J. Coulson, *J. Chem. Soc. Chem. (C),* 1968, 341.
51. J.M. Birchall, R.N. Haszeldine and A.R. Parkinson, *J. Chem. Soc.,* 1962, 4966.
52. R. Bolton and G.H. Williams, *Chem. Soc. Rev.,* 1986, 15, 261.
53. C.M. Starks and C. Liotta, *Phase Transfer Catalysis,* Academic Press, New York, 1978.
54. C.R. White, U.S. Patent 4., 642, 399 (1987) (*Chem. Abstr.,* 1987, 106, 138966t).
55. T. Papenfuhs, A. Kanschik-Conradsen and W. Pressler, Ger. Offen. DE 4,020,130 (1992) (*Chem. Abstr.,* 1992, 116, 173752p).
56. G.L. Cantrell, U.S. Patent 4,642,398 (1987) (*Chem. Abstr.,* 1987, 106, 138077u).
57. J. Yoshida, O. Furasawa and J. Kimura, *J. Fluorine Chem.,* 1991, 53, 301.
58. R.E. Banks, K.N. Mothersdale, A.E. Tipping, B.J. Cozens, and D.E.M. Wotton, *J. Fluorine Chem.,* 1990, 46, 529.
59. R.G. Pews, J.A. Gall and J.C. Little, *J. Fluorine Chem.,* 1990, 50, 365.
60. J.H. Clark and D.J. Macquarrie, *J. Fluorine Chem.,* 1987, 35, 591.
61. M. Attino, F. Cacace and A.P. Wolf, *J. Chem. Soc. Chem. Commun.,* 1983, 108.
62. J.H. Clark and N. Boechat, *Chem. Ind.,* 1991, 436.
63. D.-R. Hwang, S.M. Moerlein, L. Lang and M.J. Welch, *J. Chem. Soc. Chem. Commun.,* 1987, 1799.
64. N. Ishikawa, T. Tanabe and D. Hayashi, *Bull. Chem. Soc. Jpn.,* 1975, 48, 359.
65. R.L. Markezich, O.S. Zamek, P.E. Donahue and F.J. Williams, *J. Org. Chem.,* 1977, 42, 3435.
66. J.D. Eras, J.G. Garcia, C.A. Mathis and J.M. Gerdes, *J. Fluorine Chem.,* 1993, 63, 233.
67. A.J. Beaumont, J.H. Clark and N.A. Boechat, *J. Fluorine Chem.,* 1993, 63, 25.
68. H. Suzuki, N. Yazawa, Y. Yoshida, O. Fururawa and J. Kimura, *Bull. Chem. Soc. Jpn.,* 1990, 63, 2010.
69. M. Maggini, M. Passudetti, G. Gonzales-Trueba, M. Prato, U. Quintily and G. Scorrano, *J. Org. Chem.,* 1991, 56, 6406.
70. F. Effenberger and W. Streicher, *Chem. Berg.,* 1991, 124, 157.
71. J. Ichikawa, K.-I. Sugimoto, T. Sonada and H. Kobayashi, *Chem. Lett.,* 1987, 1985.
72. M. Van Der Puy, *J. Fluorine Chem.,* 1982, 21, 385.
73. J. March, *Advanced Organic Chemistry. Reactions, Mechanisms and Structure*, 4th ed., Wiley, New York, p. 674, 1992.
74. T.P. Simonora, V.D. Nefedov, M.A. Toropova and N.F. Kirilov, *Russ. Chem. Rev.,* 1992, 61, 584.
75. A.I. Titov and A.N. Baryshnikova, *Zh. Obsheh. Khim.,* 1953, 23, 346 (*Chem. Abstr.,* 1954, 48, 2623f).
76. D.A. Fidler, J.S. Logan and M.M. Baudakian, *J. Org. Chem.,* 1961, 26, 4014.
77. P.H. Scott, C.P. Smith, E. Kober and J.W. Churchill, *Tetrahedron Lett.,* 1970, 14, 1153.
78. T.P. Patrick, J.A. Schield and D.G. Kirchner, *J. Org. Chem.,* 1974, 39, 1758.

Chapter 3

The Balz-Schiemann Reaction and Related Chemistry

3.1 INTRODUCTION

The Balz-Schiemann reaction and related reactions are among the most important and simplest reactions used for the laboratory-scale synthesis of fluoroaromatics, and are particularly attractive since no special equipment or precautions often associated with traditional fluorination methodology are required. The reaction was developed in 1927 by Balz and Schiemann, although it is often simply referred to as the Schiemann reaction.[1] The process is a two-step route to fluoroaromatics starting from anilines, although the reaction is often preceded by aromatic nitration and reduction stages to generate the aniline precursors. The first step of the reaction involves the diazotization of an aniline (Figure 3.1). The usual Sandmeyer-type chemistry (preparation and decomposition of the aryldiazonium fluoride in aqueous solution) is unsuitable for the synthesis of fluoroaromatics, owing to the instablilty of the diazonium fluoride. Instead, the diazonium is prepared as the insoluble tetrafluroborate salt, which, unusually for a diazonium salt, is stable enough to be isolated. This step is actually the key stage in the reaction and stems from an original observation by Bart.[2] The second stage involves the thermal decomposition of the aryldiazonium fluoride to give the fluoroaromatic, typically in high yields of around 70% (Figure 3.2). Yields are highly dependent on the nature of the substrate, in particular nitro-, hydroxy-, and carboxyl-substituted aromatics generally give poor yields.

A wide number of modifications have been developed for each step of the reaction. Thus, a number of diazotization agents, including ethyl nitrite, nitrosyl, sulfuric acid, and nitrosium borotrifluoride, have been investigated. Nonaqueous solvents, including carbon tetrachloride, carbon disulfide, and ethyl acetate have been used for the diazotization step. Rather than the usual thermal decomposition, photochemical and ultrasound methods have been used. Decomposition in solvent, often with catalysis, and possibly in the actual diazotization solvent (giving a "one-pot" reaction) has also been studied.

A particularly important modification is the use of anhydrous hydrogen fluoride as the reaction solvent and source of anion for the diazonium salt, offering considerable economic advantages over the traditional tetrafluoroborate route, and avoiding the need for the isolation and transport of the intermediate diazonium tetrafluoroborate salts. Although the aryldiazonium fluorides are unstable in solution, their decomposition can be controlled to give high yields of fluoroaromatics. Although it is not readily amenable to many laboratories, the process is used extensively in

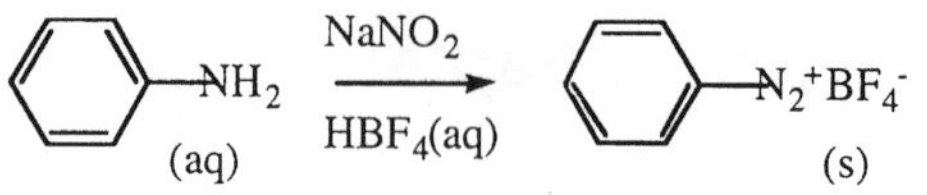

FIGURE 3.1   Preparation of aryldiazonium tetrafluoroborate salts.

FIGURE 3.2   Thermal decomposition of aryldiazonium tetrafluoroborate salts.

industry for the preparation of nonactivated fluoroaromatics, such as fluorobenzene and fluorotoluenes, and has recently been extended to continuous, rather than the usual batch, reactions. Considerable numbers of industrial processes are based on Balz-Schiemann reactions and the HF-diazotization route, with the world market for fluoroaromatics from diazotization being estimated at 25,000 tonnes, with current and planned production capacity of 50,000 tonnes.[3] In general, the Balz-Schiemann reaction is seldom employed above the 1 tonne production scale, whereas the HF-diazotization system is routinely run on the multitonne scale.

3.2   SYNTHESIS OF ARYLDIAZONIUM TETRAFLUOROBORATE SALTS

The simplest synthetic route to aryldiazonium tetrafluoroborate salts is to diazotize the aniline precursor with sodium nitrite in a 40 to 50% aqueous solution of fluoroboric acid. The yield of the diazonium tetrafluroborate is highly dependent upon the solubility in solution, and up to three mole equivalents of fluoroboric acid may be required for each mole of aniline to ensure precipitation (Figure 3.1).

A more economical strategy is to carry out the diazotization in hydrochloric or sulfuric acid to give a solution of the diazonium chloride (or sulfate). Addition of a source of tetrafluoroborate then leads to precipitation of the aryldiazonium tetrafluoroborate salt, although yields may be somewhat lower[6] (Figure 3.3). The use of sodium or ammonium tetrafluoroborate rather than fluoroboric acid as the source of tetrafluoroborate may also give an improvement in yield, since the solubility of the diazonium salt in solution increases in acidic media.[7] A reduction in the amount of dissolved diazonium tetrafluoroborate is also assisted by maintaining the volume of solvents at a minimum.

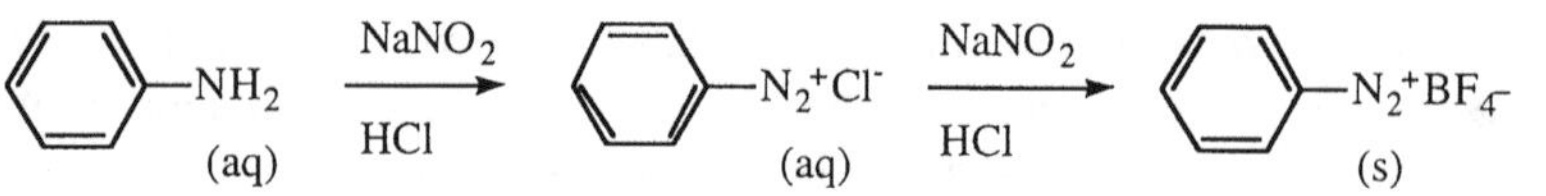

FIGURE 3.3   Synthesis of aryldiazonium tetrafluoroborate salts via diazonium chloride intermediates.

A number of reagents in addition to sodium nitrite have been used successfully for the diazotization process, including amyl nitrite,[8] boron trifluoride/hydrofluoric acid,[9] and nitrosulfuric acid in conjunction with hydroboric acid.[10]

Nonaqueous solvents such as tetrahydrafuran have also been used successfully for the preparation of aryldiazonium tetrafluoroborates and are particularly success-ful for cases in which the diazonium tetrafluoroborate is highly soluble in water, such as carboxyl- or hydroxyl-substituted anilines. Nitrosium tetrafluoroborate has been used as the diazotization agent and source of tetrafluoroborate simultaneously. The reaction has been conducted in a number of solvents, including carbon tetra-chloride, carbon disulfide, ethyl acetate, and liquid sulfur dioxide.[11] Most recently, dichloromethane has been used as the solvent, and in this case, decomposition was carried out *in situ* (Figure 3.4). The decomposition is conducted by adding a high-boiling-point, inert liquid, in this case 1,2-dichlorobenzene. Reported yields of fluoroaromatics are generally at least as good as those from traditional methods, and are significantly greater for carboxyl- and hydroxyl-substituted substrates.[12]

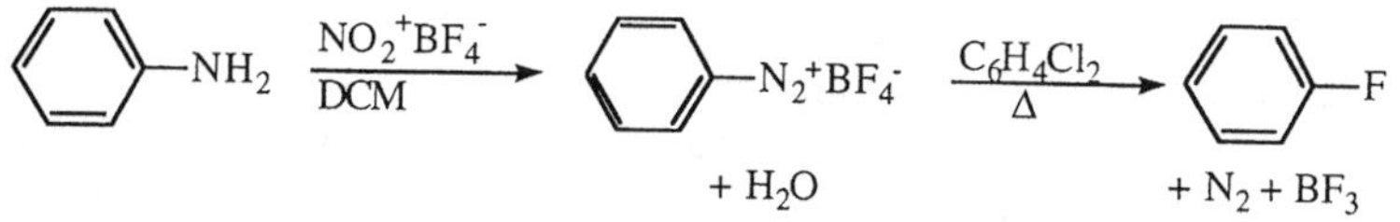

FIGURE 3.4   Diazotization using nitrosium fluoroborate and *in situ* decomposition.

Nitrosyl fluoride, generated *in situ* from an alkyl nitrate and boron trifluoride (in the form of an etherate complex), has also been used as a diazotizing reagent. The reaction is carried out in an anhydrous solvent, such as dichloromethane and the etherate complex helps to trap water generated during the reaction (Figure 3.5). Reported yields of the diazonium salt are higher than those from reactions carried out in tetrahydrofuran in some cases.[13]

FIGURE 3.5   Diazotization using nitrosyl fluoride and boron trifluoride.

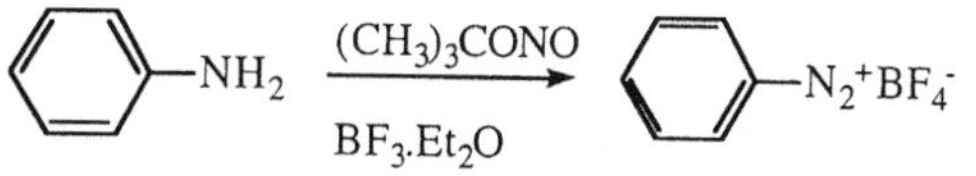

3.3 DECOMPOSITION OF ARYLDIAZONIUM TETRAFLUOROBORATE SALTS

Decomposition to the fluoroaromatic is most readily accomplished by simply heating the dry salt to decompose the complex, giving the fluoroaromatic, nitrogen, and boron trifluoride (Figure 3.6). The actual decomposition temperature is a property of each individual complex, but generally is between 100 and 200°C.[4,5] Decompo-sition is generally straightforward, with the exception of nitroaromatic diazonium tetrafluoroborates. In this case, decomposition is often violent and uncontrollable. In some cases, and usually with nitroaromatics, the rate of decomposition may be moderated by the addition of an inert diluent to the diazonium salt prior to heating.

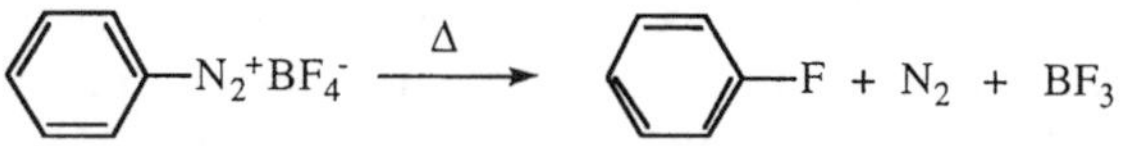

FIGURE 3.6 Thermal decomposition of aryldiazonium tetrafluoroborate salts.

Such materials include sand,[14] sodium fluoride or sodium tetrafluoroborate,[15] or barium sulfate.[16]

Decomposition need not necessarily be promoted thermally. Ultrasound, in combination with triethylamine trishydrofluoride and hydrofluoric acid resins as catalysts, has been used at room temperature, although the use of dry reagents was found to be essential.

More commonly investigated has been the photochemical decomposition of aryltetrafluoroborate salts. The standard preparation of aryldiazonium salts (reaction of the arylamine and nitrous acid in aqueous fluoroboric acid) was followed immediately with photolysis without isolation of the salt. This method has been used for the preparation of ring fluorinated tyramines and dopamines (whose nonfluorinated analogues have a role in the symantic nervous system).[18]

Photochemical decomposition has also been carried out on isolated diazonium tetrafluoroborate salts in the form of crystalline films (Figure 3.7). In this case, higher yields have been reported than from the traditional thermal decomposition process.[19] In the case of heterocyles, such as the aminopyridines, whose diazonium tetrafluoroborate salts are too unstable to be isolated, the salts wet with diethyl ether were successfully decomposed to the fluoropyridines by heating to 15–20°C in petroleum ether.[20]

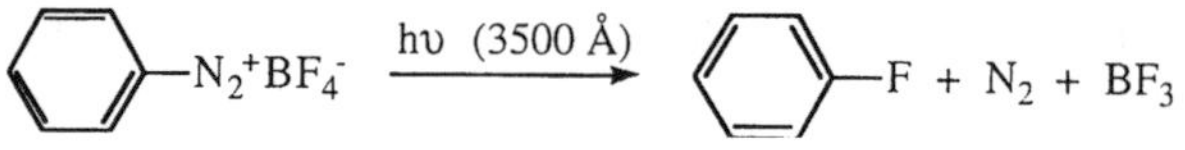

FIGURE 3.7 Photochemical decomposition of aryldiazonium tetrafluoroborate salts.

As previously discussed, addition of 1,2-dichlorobenzene to a nitrosium tetrafluoroborate–dichloromethane system allows decomposition to be carried out immediately after synthesis of the diazonium tetrafluoroborate, avoiding the need for the isolation of the intermediate salt. This would be particularly advantageous in cases where the salt is unstable. Copper powder has been used to decompose aqueous solutions of aryldiazonium tetrafluoroborates at room temperature.[21] Copper powder,[22] cuprous chloride,[22] and cupric fluoride[24] have been used to decompose solutions of isolated aryldiazonium tetrafluoroborate salts in acetone or aqueous acetone at room temperature, often with improved yields. This method was unsuitable, however, for the preparation of substituted fluoronitroaromatics,[23] or substrates containing highly polar substituents, such as nitro, hydroxyl, or carboxylate.[24]

Decomposition in solvents other than the reaction solvent may also help to moderate the reaction in cases where decomposition is difficult to control. In addition, the use of an inert moderator allows the decomposition to be carried on a larger scale than the usual thermal methods allow.[4,5]

Other solvents investigated for the decomposition of aryldiazonium tetrafluoroborate salts include petroleum ether,[25] toluene, heptane,[26] and acetonitrile dioxane.[27] 1,2-Dichloroethane has been used as a solvent for the decomposition of aryldiazonium tetrafluoroborates in the presence of crown ethers, with 21-crown-7 giving better results than 18-crown-6 and dicyclohexano-24-crown-8.[28,29]

3.4 ALTERNATIVE DIAZONIUM SALTS

Although tetrafluoroborates are the most commonly employed aryldiazonium salts, a number of other salts have been used successfully. Aryldiazonium fluorides, although too unstable to be isolated, can be converted *in situ* to fluoroaromatics, and are discussed in more detail in Section 3.5.

Anilines are converted quantitatively to their aryldiazonium hexafluorophosphate salts by treatment with hexafluorophosphoric acid (Figure 3.8). Thermal decomposition of the salt often gives improved yields of the fluoroaromatic products.[31] Since phosphorus pentafluoride is a weaker Lewis acid than boron trifluoride, side reactions may be expected to be reduced using this system. Diazonium hexafluorophosphates have recently been used to prepare a number of biologically active fluoroaromatics. A key step in the synthesis of Enoxacin, a new antibacterial pyridonecarboxylic acid was the Schiemann reaction of a aminopyridine which was successfully accomplished via the diazonium hexafluorophosphate. A number of routes for the decomposition were investigated, including heating with magnesium sulfate, and thermal decomposition in solvents, including xylene, cyclohexane, carbon tetrachloride, ethyl acetate, isopropyl acetate, heptane, and toluene.[32]

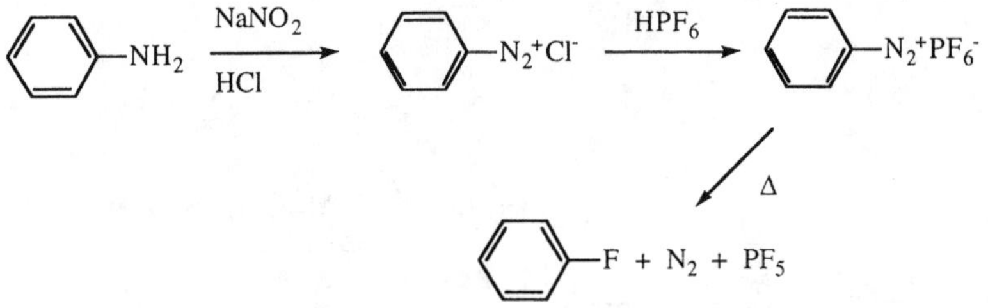

FIGURE 3.8 Synthesis and decomposition of diazonium hexafluorophosphates.

Enoaxicin

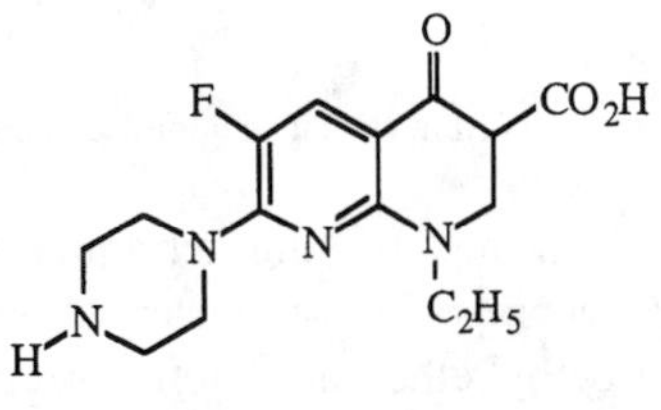

A series of 7-amino-1-cyclopropyl-8-fluoro-1,4-dihydro-4-oxo-1,1-naphthyyridine-3-carboxylic acids as potential antibacterial agents were prepared, in which a key stage again used this modification to introduce a fluorine atom into an intermediate pyridine.[33]

Treatment of anilines with fluorosilicic acid and a diazotizing agent such as ethyl nitrite generates the diazonium hexafluorosilicate, which can then be isolated and decomposed to give the fluoroaromatic. This system has been used for the preparation of a number of fluoroaromatics unobtainable from conventional Balz-Schiemann routes.[30] Hexafluoroantimonate salts are generally unsuitable for the synthesis of fluoroaromatics, which is not unexpected, since the antimony pentafluoride liberated would be expected to react with the fluoroaromatic product even at low temperatures.[34]

3.5 MECHANISTIC ASPECTS OF THE BALZ-SCHIEMANN REACTION

The mechanism of the reaction is not fully understood, but it appears that a number of competitive processes may occur (Figure 3.9). From the decomposition of 4-(t-butyl)benzenediazonium tetrafluoroborate in dichloromethane, the rate-determining step has been determined to be the decomposition of the aryldiazonium tetrafluoroborate, to give the phenyl cation, nitrogen, and boron trifluoride, rather than direct attack by a fluoride source.[35] Addition of bromine gave no bromoaromatic products, which would be expected from a radical-type mechanism. Addition of excess boron trifluoride, (altering the equilibrium $BF_4^- \rightleftharpoons BF_3 + F^-$) had no effect on the rate of reaction, from which the source of fluoride was determined to be the tetrafluoroborate, rather than the fluoride anion.[36]

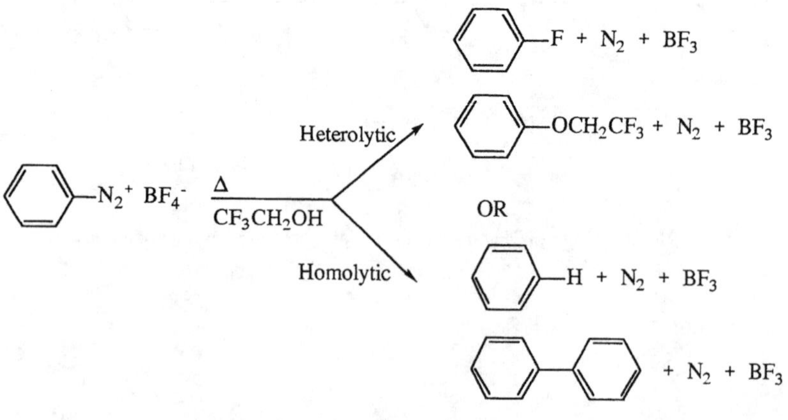

FIGURE 3.9   Reaction products from aryldiazonium decomposition in trifluoroethanol.

The decomposition of benzenediazonium tetrafluoroborate has been studied in a number of solvents. The decomposition was carried out in 2,2,2-trifluoroethanol, giving phenyl 2,2,2,-trifluoroethyl ether and fluorobenzene as products. Addition of increasing amounts of pyridine to this system results in the increase of homolytic-type reaction products, including benzene, biphenyl, and diazo-tars.[37]

The decomposition of aryldiazonium tetrafluoroborates, tetrachloroborates, and tetrabromoborates in fluorobenzene produced, in addition to the haloaromatic product, *ortho-* and *para-*fluorobiphenyls, suggesting an electrophilic-type arylation.[38] The decomposition in nitrobenzene gave the m-isomer of nitrobiphenyl only, suggesting a radical-type substitution,[39] although addition of a copper catalyst to the system resulted in formation of the ortho and para isomers.[40] Radical-type mechanisms have recently been postulated based on a mass spectroscopic investigation.[41]

3.6 FLUOROAROMATICS VIA HF-BASED DIAZOTIZATIONS

Aromatic amines may also be converted to the corresponding fluoroaromatics by diazotization in anhydrous HF, in a reaction closely related to the Balz-Schiemann reaction.[42] The intermediate aryldiazonium fluoride is unstable, and is decomposed *in situ* to the fluoroaromatic by heating (Figure 3.10). The reaction is widely used industrially, and is normally the method of choice for nonactivated aromatics, such as fluorobenzene and fluorotoluenes.[3] The process has obvious advantages over traditional Balz-Schiemann methodology — the one-pot reaction does not require the isolation of intermediate salts, and the process does not require the use of tetrafluoroborate salts, which results in lower production costs.

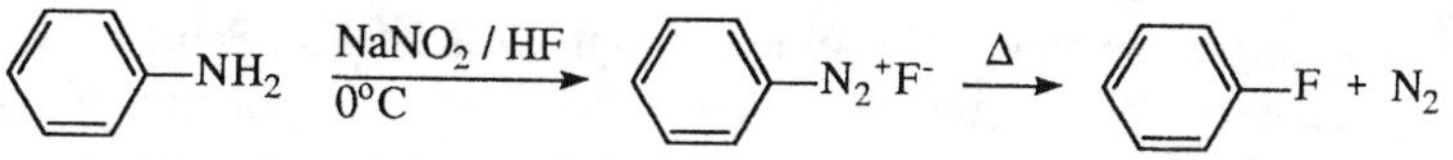

FIGURE 3.10 Aniline diazotization in anhydrous hydrogen fluoride.

Recent developments of this method include the use of continuous, rather than the usual batch-type, processes for preparation[43] and decomposition of the aryldiazonium fluorides.[44,45] Suitable diazotizing agents include sodium nitrite,[42] nitrosyl fluoride-hydrogen fluoride,[44] and nitrosyl chloride-hydrogen fluoride.[46] The reaction is carried out by addition of the diazotization reagent to a solution of the aniline in anhydrous hydrogen fluoride at around 0°C. After allowing the diazotization to proceed, the mixture is warmed to decompose the diazonium fluoride and then heated to distill out the fluoroaromatic. The procedure is limited by the temperature required to decompose the salt, which is limited by the highest temperature achievable in anhydrous hydrogen fluoride. The addition of an additional component may increase the maximum temperature achievable. Such components include ammonium fluoride,[47] ammonium bifluoride,[47] sodium fluoride, water, or glyme.[49] The reaction has also been carried out in inert solvents such as $MeOCH_2CH_2OMe$.[50]

Substituents on the aromatic amine have a number of effects on the success of the reaction. The rate of decomposition is also slowed down for substrates in which the charge on the diazonium group can be delocalized onto another substituent (e.g., ortho or para nitro). Ortho groups such as amino, methoxy, nitro, and chloro may prevent decomposition of the diazonium fluoride altogether by stabilization resulting from ortho lone-pair delocalization.[42,49] Groups such as methoxy, acetyl, and methyl may react with the diazotization reagents. Electron donating groups (e.g., methyl) increase the rate of decomposition,[49] whereas electron withdrawing groups such as nitro, hydroxyl, and carbonyl decrease the rate.[49,42]

The diazotization reaction has also been carried out in anhydrous hydrogen fluoride in the presence of pyridines. Fluorination by pyridinium poly(hydrogen fluoride), a "tamed" form of HF, gives reasonable yields of fluoroaromatics from anilines at room temperature, although mixtures of isomers are sometimes obtained[51] (Figure 3.11). The yield of fluoroaromatic can often be increased by the heating of the aryldiazonium fluoride without volatile HF being present. This is most often accomplished by addition of an organic base to the diazotization reaction, such as pyridine, pyrazine, or 2-hydroxypyridine.[52] The procedure has also been applied to the synthesis of fluoropyridines, yields from which are often low via traditional Balz-Schiemann methods.[53] Photochemical, rather than thermal, decomposition has been reported to give higher yields in some cases.[54] Yields of the fluoroaromatic are also affected by the amount of base present, and the HF/aniline ratio.[55]

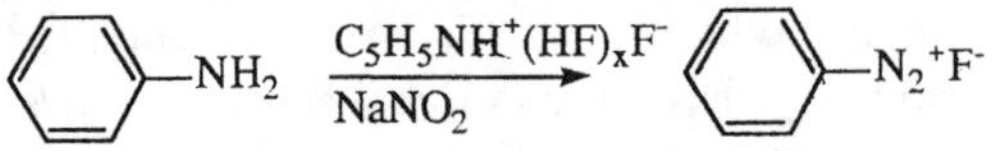

$$\text{Ph-NH}_2 \xrightarrow[\text{NaNO}_2]{\text{C}_5\text{H}_5\text{NH}^+(\text{HF})_x\text{F}^-} \text{Ph-N}_2^+\text{F}^-$$

FIGURE 3.11   Fluorination by pyridinium poly(hydrogen fluoride).

3.7 SYNTHETIC METHODS

For the laboratory-scale preparation of fluoroaromatics, the Balz-Schiemann reaction is one of the most readily accessible methods available, and is undoubtably the most commonly employed; experimental details for a number of modifications of the reaction are well documented.[4,5] The reaction can usually be carried out in standard laboratory equipment, with few of the precautions normally associated with fluorine chemistry being necessary.

The Balz-Schiemann reaction does have a number of significant disadvantages:

Yields are often not reproducible.

The success of the reaction is greatly influenced by other ring substituents. Ortho-substituted amines generally give lower yields than the other isomers.[4,5] The synthesis of ortho-fluoronitroaromatics is particularly difficult — the use of halogen exchange routes may be preferred in this situation. Sulfonic acid groups interfere with the diazotization procedure; esters, carboxyl, amino, nitro, and hydroxyl substituents generally decrease yields in that order.[4,5]

Side reactions during decomposition may reduce yields and product purity. These may include reaction with the liberated boron trifluoride, and radical-type reactions. Typically, these products will include the reduced product (replacement by hydrogen rather than halogen), coupled products (biphenyls) and "diazo tar", a complex mixture of polyaromatics. In some cases, unidentifiable "diazo tars" components may be the major reaction product. Chloro- rather than fluoroaromatics may result from contamination by diazonium chloride salts present as impurities.[57] Fluorine groups ortho to the diazo group are also prone to replacement by chloride during the reaction.[58]

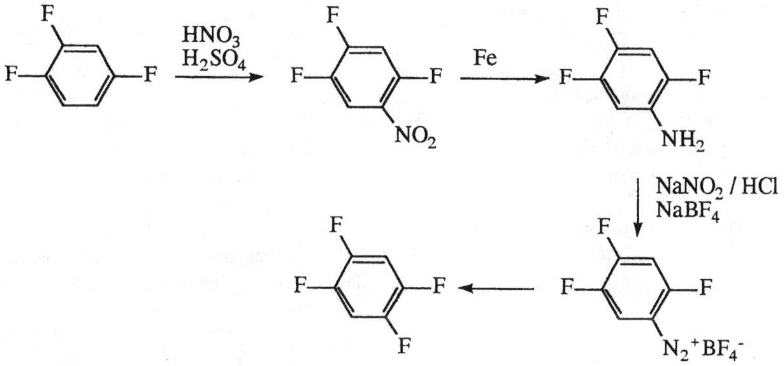

FIGURE 3.12 Sequential nitration, reduction, and diazotization route to multifluoroaromatics.

The Balz-Schiemann reaction can be used to introduce two fluorine atoms simultaneously into a molecule, although yields are reduced if the two amine groups are present on the same aromatic ring. Sequential nitration, reduction, and diazotization has been used for the preparation of tetrafluorobenzene[56] (Figure 3.12).

Table 3.1 summarizes the most important routes for the synthesis of aryldiazonium salts. The simplest method is simply to diazotize the aqueous solution of the amine in borofluoric acid with sodium nitrite. More forcing conditions may be required for weakly basic amines, such as the use of a more powerful diazotizing agent. Highly soluble amines may be better diazotized in hydrochloric or sulfuric acid, followed by addition of a source of tetrafluoroborate, or preferably in an organic solvent such as dichloromethane, carbon tetrachloride, or tetrahydrofuran. The preparation of alternative diazonium complexes, particularly the diazonium hexafluorophosphates, gives improved yields in a number of cases.

The diazonium tetrafluoroborate salt is precipitated during the course of the diazotization. For unstable diazonium salts, decomposition without isolation may be the preferred method. *In situ* decomposition, typically by catalysis using a copper salt, avoids the need to isolate the salt. Isolation of the diazonium salt typically gives yields of up to 90%. The isolation allows further purification, either by simply washing or, if the complex is sufficiently stable, by recrystallization.

Decomposition methods are summarized in Table 3.2. Decomposition is usually effected simply by heating the aryldiazonium salt.[4,5] Heating the salt under a stream of inert gas assists in the removal of boron trifluoride, which helps to prevent reaction with the fluoroaromatic product. In cases where the compound does not decompose smoothly (nitroaryldiazonium decompositions are generally very violent), the decomposition can be assisted by the use of an inert bulk material, such as sand, sodium fluoride, or sodium tetrafluoroborate. Metal chloride salts should be avoided to prevent formation of chloroaromatic products. Alternatively, decomposition may be carried out in an inert solvent such as toluene. This helps to control the temperature of the process and allows the scale of the reaction to be increased.

Photochemical decomposition has been reported to give increased yields in several cases, either in solution or as crystalline films, although the method is more

Table 3.1 Synthetic Routes to Aryldiazonium Salts

Method	Comments
Diazotization of amine in borofluoric acid	Simplest procedure
Diazotization of amine in hydrochloric acid; addition of $HBF_4/NaBF_4/NH_4BF_4$	Less expensive, yields may be increased, but product may be contaminated with chloroaromatics
Diazotization in nonaqueous solvent	Useful for highly soluble diazonium complexes
Preparation of diazoniumhexafluorophosphate	Improved yields in some cases
Diazotization/decomposition in AHF	Requires special handling precautions/equipment; suitable for many processes where traditional Balz-Schiemann fails

Table 3.2 Decomposition Methods for Aryldiazonium Salts

Method	Comments
Decomposition *in situ*	Simplest procedure, isolation of salt not required; Cu salts catalyze reaction in aqueous systems; dichloromethane solutions of diazoniums may be decomposed *in situ*
Isolation of salt	Additional stage, but allows purification before decomposition
Dry decomposition	Simple procedure, may require moderator (sand, salt) for some aryldiazoniums
Decomposition in solution	Moderates reaction, allows increase in scale of operation
Photochemical decomposition	Improved yields in some cases

complex than simple thermal decomposition. However, the scale on which the latter method can be conducted is much lower than many of the other methods.

Diazotization in anhydrous hydrogen fluoride is often successful when the traditional Balz-Schiemann reaction fails, or gives poor yields. However, the use of anhydrous hydrogen fluoride is not trivial, and is impractical for many traditional organic chemistry laboratories. In addition, special reaction vessels are required for the handling and control of reactions involving anhydrous hydrogen fluoride.

REFERENCES

1. G. Balz and G. Schiemann, *Chem. Ber.,* 1927, 60, 1186.
2. H. Bart, German Patent 281,055 (1915).
3. L. Dolby-Glover, *Chem. Ind. (London),* 1986, 518.
4. A. Roe, *Org. React.,* 1949, 5, 193.
5. H. Suschitzky, *Adv. Fluorine Chem.,* 1965, 4, 1.
6. E.B. Starkey, *J. Am. Chem. Soc.,* 1946, 68, 793.
7. U.S. Patent 1,961,327 (1933).
8. *J. Org. Chem.,* 1949, 328.
9. U.S. Patent 1,961, 327 CA 27, 4539, 1933.
10. A.P.J. Luttringhaus and H. Neresheimer, *Liebigs. Ann.,* 1929, 259.
11. U. Wannagat and G. Hohlstein, *Chem. Ber.* 1955, 88, 1839.
12. D.J. Milner, *Synth. Commun.,* 1988, 22, 73.
13. M.P. Doyle and W.J. Bryker, *J. Org. Chem.,* 1979, 44, 1572.
14. G. Schiemann and R. Pillarsky, *Chem. Ber.* 1929, 60, 3035.

15. A. Roe and H.C. Fleichmann, *J. Am. Chem. Soc.,* 1947, 69, 509.
16. R.J.W. Lefevre and E.E. Turner, *J. Chem. Soc.,* 1930, 1158.
17. A. Muller, U. Roth and R. Miethchen, *J. Fluorine Chem.,* 1985, 29, 205.
18. K.L. Kirk, *J. Org. Chem.,* 1976, 41, 2373.
19. R.C. Peterson, A. DiMagggio, A.L. Herbert, T.J. Haley, J.P. Mykytka and I.M. Sarker, *J. Org. Chem.,* 1971, 36, 631.
20. A. Roe and G.F. Hawkins, *J. Am. Chem. Soc.,* 1947, 69, 2443.
21. E.D. Bergmann and M. Bentov, *J. Org. Chem.,* 1954, 19, 1594.
22. E.D. Bergmann, S. Berkovic and R. Ikar, *J. Am. Chem. Soc.,* 1956, 78, 603.
23. I.K. Barben and H. Suschitzky, *Chem. Ind. (London),* 1957, 1693.
24. E.D. Bergmann and S. Berkovic, *J. Org. Chem.,* 1961, 26, 919.
25. G. Schiemann, *J. Prakt. Chem.,* 1934, 140, 97.
26. B.L. Zenitz and W.H. Hartung, *J. Org. Chem.,* 1946, 11, 444.
27. M.A. Goldberg, E.P. Ordas and G. Carsch, *J. Am. Chem. Soc.,* 1947, 69, 260.
28. H. Nakazumi, I. Szele and H. Zollinger, *Tetrahedron Lett.,* 1981, 22, 3053.
29. H. Nakazumi, I. Szele, K. Yoshida and H. Zollinger, *Helv. Chim. Acta,* 1983, 66, 1721.
30. P.H. Cheek, R.H. Wiley and A. Roe, *J. Am. Chem. Soc.,* 1949, 71, 1863.
31. K.G. Rutherford, W. Redmond and J. Rigamonti, *J. Org. Chem.,* 1961, 26, 5149.
32. J. Matsumoto, T. Miyamoto, A. Minamida, Y. Nishimura, H. Egawa and N. Nishimura, *J. Heterocyclic Chem.,* 1984, 21, 673.
33. J.P. Sanchez and R.D. Gogliotti, *J. Heterocyclic Chem.,* 1993, 30, 855.
34. W. Lange and K. Askitopoulos, *Z. Anorg. Chem.,* 1935, 223, 369.
35. C.G. Swain, J.E. Sheats and K.G. Harbison, *J. Am. Chem. Soc.,* 1975, 97, 796.
36. C.G. Swain and R.J. Rogers, *J. Am. Chem. Soc.,* 1975, 97, 799.
37. P. Burri, H. Loewenschuss, H. Zollinger and G.K. Zwolinski, *Helv. Chim. Acta,* 1974, 57, 395.
38. G.A. Olah and W.S. Tolgyesi, *J. Org. Chem.,* 1961, 26, 2053.
39. A.N. Nesmeyanov, L.G. Makarova and T.P. Tolystaya, *Tetrahedron,* 1957, 1, 145.
40. L.G. Makarova and M.K. Matveena, *Izv. Akad. Nauk. SSSR. Otdel Chim. Nauk.,* 1960, 1974.
41. Y. Deng, *Acta. Chim. Sin. (Engl. Ed.),* 1989, 422.
42. R.L. Ferm and C.A. VanderWerf, *J. Am. Chem. Soc.,* 1950, 72, 4809.
43. N.J. Stepaniuk and B.J. Lamb, U.S. Patent 124,500 (1987).
44. N.J. Stepaniuk and B.J. Lamb, U.S. Patent 124,501 (1987).
45. Y. Yoda, H. Hokogahara, I. Harada and T. Kuroda, Japanese Patent 03,167,143 (1991).
46. W. Shenk and G.R. Pellon, U.S. Patent 2,563,796 (1951).
47. M.M. Boudakain, U.S. Patent 7,07,975, (1976).
48. M.M. Boudakain, U.S. Patent 4,075,252, (1978).
49. J.S. Moillet, *J. Fluorine Chem.,* 1987, 35, 38.
50. J.S. Moillet, U.K. Patent 2,173,188. 107, 23064r.
51. G.A. Olah, J.T. Welch, Y.D. Vankar, M. Nojima, I. Kerekes and J.A. Olah, *J. Org. Chem.,* 1979, 44, 3872.
52. T. Fukuhara, N. Yoneda, T. Sawada and A. Suzuki, *Synth. Commun.,* 1987, 17, 685.
53. T. Fukuhara, N. Yoneda, and A. Suzuki, *J. Fluorine Chem.,* 1988, 38, 435.
54. N. Yoneda, T. Fukuhara, T. Kikuchi and A. Suzuki, *Synth. Commun.,* 1989, 19, 865.
55. T. Fukuhara, S. Sasaki, N. Yoneda and A. Suzuki, *Bull. Chem. Soc. Jpn.,* 1990, 63, 2058.
56. G.C. Finger, F.H. Reed, D.M. Burness, D.M. Fort and R.R. Blough, *J. Am. Chem. Soc.,* 1951, 73, 145.
57. H. Suschitzky, *J. Chem. Soc.,* 1953, 3042.
58. G.C. Finger and R.E. Oesterling, *J. Am. Chem. Soc.,* 1956, 78, 2593.

Chapter 4

Other Aromatic Fluorination Methodologies

4.1 INTRODUCTION

The Balz-Schiemann and halex reactions discussed in previous chapters are undoubtedly the most common routes used in the synthesis of fluoroaromatics on both the laboratory and industrial scale. However, these techniques do have significant drawbacks. The Balz-Schiemann reaction requires an aniline precursor, which, in turn, requires a nitro aromatic precursor. The halex methodology requires chloro-substituted substrates, and usually activation of the chloro group towards nucleophilic substitution. In both cases, other functional groups present on the ring can cause significant side reactions or otherwise adversely affect the success of the fluorination.

Direct fluorination would therefore offer significant advantages over either of the two established routes. The process would not require particular functional groups (other than hydrogen) for the replacement. Although the direct introduction of other halogens into aromatic rings is well established, fluorination with elemental fluorine is nontrivial, owing to the extremely reactive nature of the element and its toxicity and corrosive properties. Until recently, the methodology associated with the direct fluorination of aromatics has been limited to those experienced in the handling of elemental fluorine (either as the fluorinating agent itself, or as a reagent in the synthesis of another fluorinating compound) or anhydrous hydrogen fluoride. In recent years, a number of commercially available sources of electrophilic fluorine have become available for use on both a laboratory and industrial scale.

^{18}F radiochemical applications have become increasingly important in recent years, for example for use in positron emitting transaxial tomography (PETT) scanning. Since the half life of ^{18}F is relatively short (approximately 120 min), it is essential that fluorination be carried out as quickly as possible. Many of the reagents described in this chapter can effect fluorination in minutes under particularly mild conditions. Since the majority of these reagents are derived from elemental fluorine itself, the radiochemical yield is also much higher than, for example the Balz-Schiemann reaction, where only 25% of the $[^{18}F] BF_4^-$ would actually be incorporated into the fluoroaromatic product.

Both the halex-type reactions and Balz-Schiemann reactions operate via nucleophilic substitution routes, which means that the accessibility of fluoroaromatics is often limited to certain isomers. Since the reagents discussed in this chapter operate via an electrophilic mechanism, there is considerable scope for the preparation of isomers not normally accessible via the other routes, or where other routes simply give poor yields.

69

A number of reagents have been investigated for applications in the direct fluorination of aromatics, or for the electrophilic replacement at carbanionic aromatic centers such as aryl Grignards. Elemental fluorine itself has been used, although the reaction needs to be moderated by the use of diluted fluorine in inert solvents. Other fluorinating reagents are generally prepared from elemental fluorine, resulting in less reactive reagents and more selective fluorinations. Reagents of this class include high-valency metal fluorides (such as cobalt trifluoride and silver difluoride), xenon difluoride, trifluoromethyl hypofluorite, acetyl hypofluorite, and cesium fluoroxysulfate. Recently, a number of electrophilic fluorinating reagents, based on a N-F$^+$ system have been developed for the selective fluorination of both aromatic and aliphatic systems. Although these reagents are derived from elemental fluorine, a number are commercially available, which has opened up the field of electrophilic fluorination to workers not practiced in the handling of the element.

4.2 DIRECT FLUORINATION

Traditionally, the direct fluorination of organic compounds has been considered an uncontrollable process, with little, if any selectivity. This is primarily due to the weak F-F bond in fluorine itself, compared to the relatively strong C-F bond, providing a strong thermodynamic driving force for organic fluorination. In aromatic systems, perfluorination, often accompanied by ring-opening fluorinating, is common. Explosive reactions have been reported in many cases. Recently, methods have been developed for the selective fluorination of aromatic compounds with elemental fluorine by "taming" the highly reactive element.

Fluorination by atomic, rather than molecular fluorine is possible by the use of a radio frequency plasma discharge.[1,2] In this case, selective fluorination is possible where previously, without the discharge, only polymeric tars resulted. The fluorination is selective towards the para, rather than the ortho or meta position. The reaction is believed to follow an electrophilic substitution mechanism via Π complexes, with the fluorine intermediate interacting with the entire Π electron system, rather than being bonded to any particular carbon atom. In the fluorination of benzotrifluoride, almost entirely para fluorination results with a fluorine-to-benzotrifluoride ratio greater than 10:1, which is believed to be the result of steric effects.[3]

A more practical method for the direct fluorination of aromatics is to dilute the fluorine with an inert gas such as nitrogen or argon, and to carry out the fluorination in an inert solvent such as fluorotrichloromethane, acetonitrile, or hexafluorobenzene. Using this system at very low levels of conversion (<0.01%), the Hammett ρ^+ value for F_2 is –2.45, which is indicative of a highly reactive, unselective reagent, operating via a polar, electrophilic substitution mechanism.[4] This system has also been developed for use on a molar scale. The substrate is used as a 6% solution in acetonitrile, and the reaction conducted at –35°C, with a 0.7:1 ratio of fluorine to substrate. Higher fluorine ratios give complex tar-like products, composed primarily of highly fluorinated polycyclohexanes. In the fluorination of benzene, fluorobenzene and ortho-, meta-, and para-difluorobenzene (60:4:1:5 ratio) result (Figure 4.1).

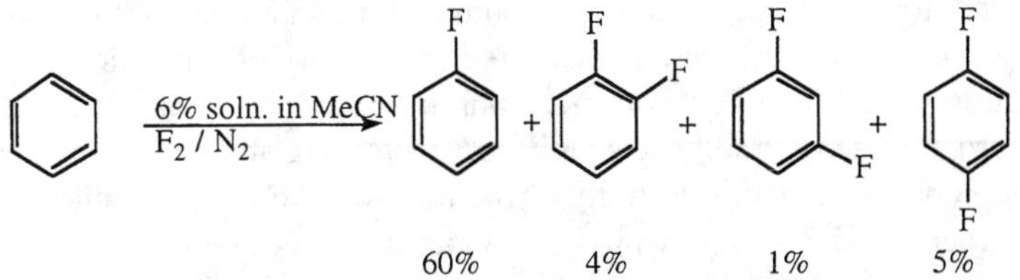

FIGURE 4.1   Direct fluorination of benzene using elemental fluorine.

Again, an ionic electrophilic substitution mechanism operates. Toluene is also readily fluorinated at –70°C, but methyl benzoate also undergoes fluorination on the methyl group as well as on the ring.[5]

Aryl oxygen-substituted aromatics (including phenols and cresols,[6] hydroxybenzoic acids, hydroxyaldehydes, phenyl salicylate, and biphenyl derivatives[7]) have also been directly fluorinated with elemental fluorine. High ortho regioselectivity is observed with low levels of conversion (50%); at higher conversion levels, the fluorinated product converted to polymeric materials. A number of solvents, including acetonitrile, tetraglyme, methanol, and chloroform, have been studied. In the fluorination of phenol, anhydrous hydrogen fluoride has been studied as a solvent, although poorer levels of conversion result, the para–ortho ratio is 1:1, unlike the 1:2 ratio observed in acetonitrile.

Most recently, selective, direct fluorination of aromatic substrates has been reported using protic acids (formic or sulfuric acid).[8] The reactions may be carried out at room temperature, in contrast to the low temperatures (–25°C to –78°C) often employed for similar fluorinations using a 10% F_2/N_2 mixture. The role of the acid may be either as a strong acid catalyst, or as part of an *in situ* formation of a hypofluorite, which then rapidly reacts with the substrate.

Substituted pyridines have been prepared by fluorination in $CF_2ClCFCl_2$ at –25 to 25°C, using 10% F_2 in N_2. At temperatures below -25°C, however, potentially explosive difluorides may be formed (Figure 4.2). This route is particularly useful for 2-fluoropyridines, which are not normally accessible via Balz-Schiemann or halex-type chemistry.[9]

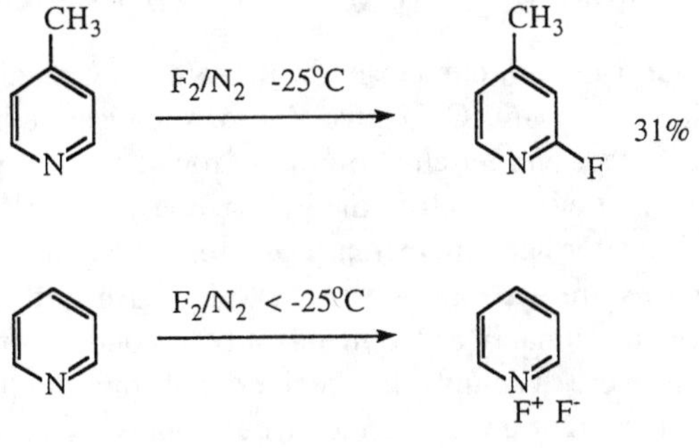

FIGURE 4.2   Fluorination of substituted pyridines using elemental fluorine.

The addition of a Lewis acid, such as aluminum trichloride (or preferably boron trichloride) often improves the yield and regioselectivity of aromatic fluorinations. The addition of a 1 mole equivalent of boron trichloride to nitrobenzene allows direct fluorination to be readily accomplished. Increasing amounts of the Lewis acid generally increase the para selectivity of the reaction. Activated aromatics, such as toluene, generally give complex mixtures under these conditions.[10]

Molecular sieves have been used to increase yields of fluoroaromatics. The substrate is adsorbed onto a bed of molecular sieves, which limits contact between substrate radicals during the fluorination reaction. The elemental fluorine is passed over the impregnated bed at $-78°C$, and increasing the concentration of fluorine during the course of the reaction improves yields.[11]

In addition to the direct fluorination of hydrocarbons, elemental fluorine has also been used to replace metallic and silyl ligands on aromatic rings. The reaction of aryl tin complexes with ^{18}F-labeled fluorine yields ^{18}F-labeled aromatics suitable for radiochemical applications. Yields from the use of fluorine are much more variable than those from the reaction with acetyl hypofluorite. ^{18}F-labeled fluoroaromatics have also been prepared by the reaction of aryl silanes with labeled fluorine. This reaction is very rapid (8 min), and is therefore very suitable for radiochemical synthesis. Yields are generally highest for trialkyl-substituted arylsilanes, and are reduced somewhat when the ring contains electron-withdrawing groups[12] (Figure 4.3). The fluorination of aryl trimethylsilanes has been reported to give ring fluorination at other positions, in addition to replacement of the silyl group, although the yields are strongly dependent upon reaction temperature and substrate concentration.[13]

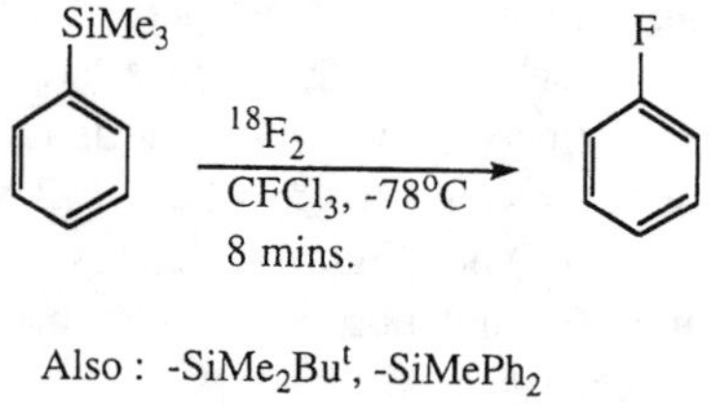

$$SiMe_3 \xrightarrow[\substack{CFCl_3, -78°C \\ 8 \text{ mins.}}]{^{18}F_2} F$$

Also : -SiMe$_2$Bu^t, -SiMePh$_2$

FIGURE 4.3 Conversion of aryl trimethylsilyls to fluoroaromatics using elemental fluorine.

Grignard, and aryl lithium complexes, which are more readily available, react with elemental fluorine at $-60°C$ to give fluoroaromatics. A hydrocarbon ether solvent is generally used, and tetrahydrofuran is found to be a poorer solvent for the reaction than dialkyl ethers such as diethyl or dibutyl ether.[14] The direct fluorination of aryl tin, lead, germanium, mercury, and thallium complexes has also been investigated.[15,16] Yields are generally observed to increase further down the group in the periodic table, presumably as the metal-carbon, bond becomes weaker. Since mercury complexes were insoluble in the inert solvents studied (fluorotrichloromethane, carbon tetrachloride, and acetonitrile), yields were low; in the case of aryl thallium salts, no conversion to the fluoroaromatic was observed. Aryl thallium complexes have been successfully fluorinated by the use of aryl thallium (III)

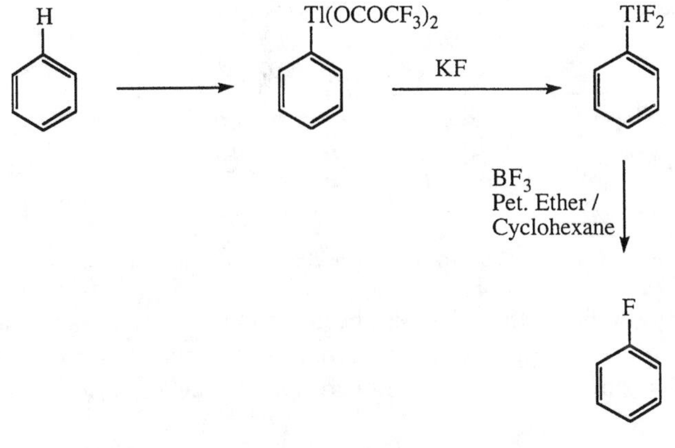

FIGURE 4.4 Synthesis and fluorination of aryl thallium compounds.

difluorides. Reasonable yields are obtained by bubbling boron trifluoride through a suspension of the salt in petroleum ether or cyclohexane, possibly via tetrafluoroborate intermediates (Figure 4.4).

4.3 FLUORINATION USING O-F-TYPE COMPOUNDS

4.3.1 Fluoroxytrifluoromethane and Bis(fluoroxy)trifluoromethane

Fluoroxytrifluoromethane (trifluoromethyl hypofluorite) is prepared by the direct fluorination of carbon monoxide with a cesium fluoride catalyst for the addition of a second mole equivalent of fluorine. Similarly, bis(fluoroxy)trifluoromethane is prepared from carbon dioxide[17] (Figure 4.5). Although the reagents require the use of elemental fluorine for their synthesis, they are relatively safe to handle once formed. Fluoroxytrifluoromethane will directly fluorinate benzene, with a small amount of benzenetrifluoromethyl ether as a side product. Toluene and xylenes generally give poor yields and complex mixtures, and the reagents are unsuitable for deactivated aromatics such as nitrobenzene. Reasonable yields have been reported, however, for anisoles and cresols. 2-Fluoro-4-trifluoromethylaniline has been synthesized using fluoroxytrifluoromethane via the acetanilide derivative (Figure 4.6). The reagents often show a strong ortho selectivity when fluorination is carried out in nonpolar aprotic solvents.

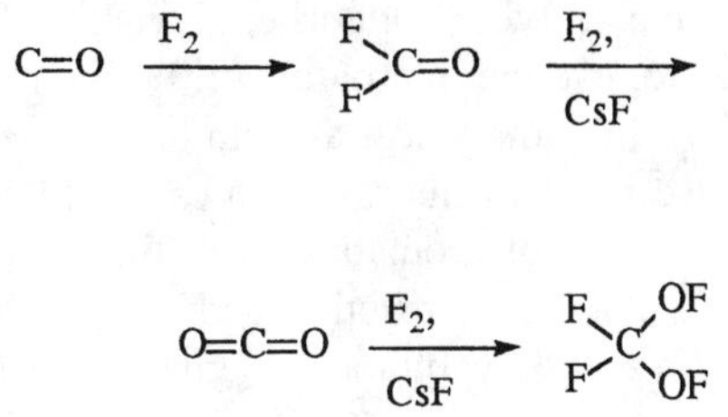

FIGURE 4.5 Synthesis of fluoroxytrifluoromethane and bis(fluoroxy)trifluoromethane.

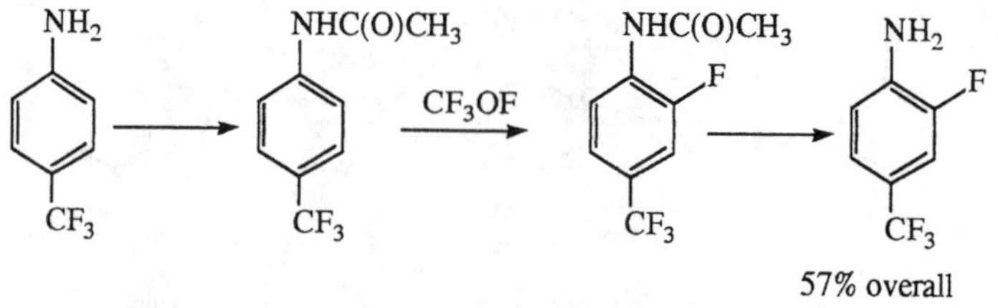

57% overall

FIGURE 4.6   Synthesis of 2-fluoro-4-trifluoromethylaniline using trifluoromethyl hypofluorite.

Fluoroxytrifluoromethane has also been used to convert aryl tin or aryl mercury compounds to the corresponding fluoroaromatic.[18] Diphenylmercury gives an 83% yield of fluorobenzene, although conversions are lower for aryl tin compounds.

4.3.2 Acetyl Hypofluorite

Acetyl hypofluorite may be prepared by bubbling elemental fluorine through a suspension of sodium acetate or sodium fluoride in a mixture of fluorotrichloromethane and acetic acid at −75°C.[19] Aromatics react to give ortho- and para-substituted products, although often the reaction is only successful for activated substrates.[20] The reaction mechanism for aryl oxygen compounds is believed to proceed via a 1,2 addition across the high Π electron density region between the ipso and ortho positions on the ring, followed by an elimination reaction (Figure 4.7).

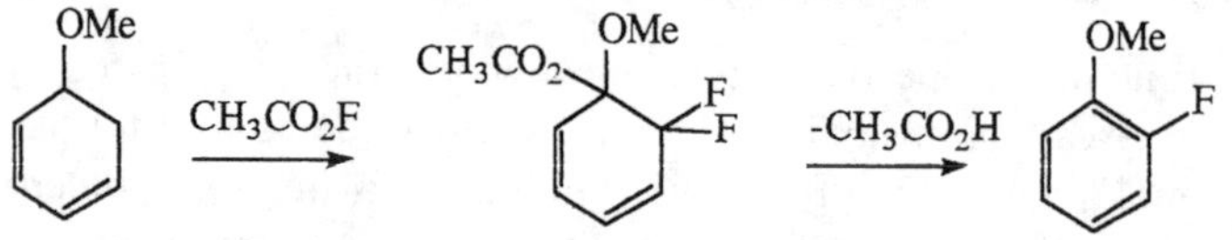

FIGURE 4.7   Mechanism for the fluorination of aryl oxygen compounds using acetyl hypofluorite.

The reagent is much milder in fluorination than trifluoromethyl hypofluorite, although reaction is still rapid. This is particularly important in ^{18}F radio labeling cases, where rapid synthesis of the fluoroaromatic is essential. ^{18}F-substituted aromatics have been prepared from ^{18}F-labeled acetyl hypofluorite by the reaction with a number of aryl precursors, including the potassium salts of aryl pentafluorosilicates,[21] aryl mercury, and aryl tin compounds. Aryl mercury compounds, particularly aryl mercury acetates, react at room temperature in under 5 min to give fluoroaromatics.[22] The mechanism has been postulated to involve oxidation of the aromatic ring by the acetyl hypofluorite, giving a radical cation, which undergoes attack by fluoride to give the fluoroaromatic[23] (Figure 4.8).

Aryl tin compounds, as with aryl mercury or pentafluorosilicates offer a regioselective route to fluoroaromatics, avoiding the isomers often obtained with the direct reaction of aromatics. Aryl tin derivatives are readily prepared from their bromo analogues by reaction with n-butyl lithium followed by a tin complex, and react with acetyl hypofluorite to give fluoroaromatics in approximately 70% yields[24] (Figure 4.9).

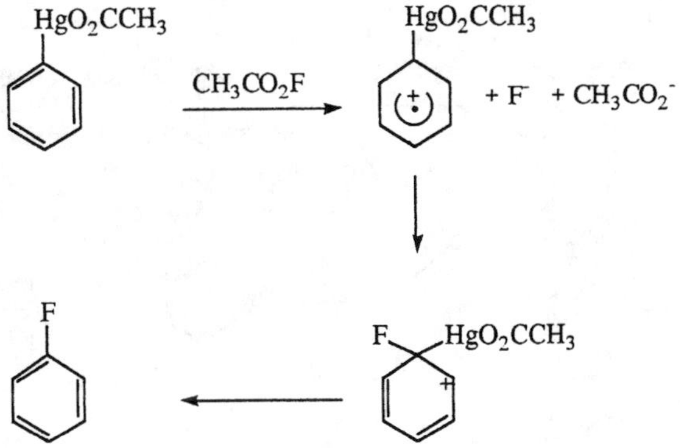

FIGURE 4.8 Fluorination of aryl mercury compounds using acetyl hypofluorite.

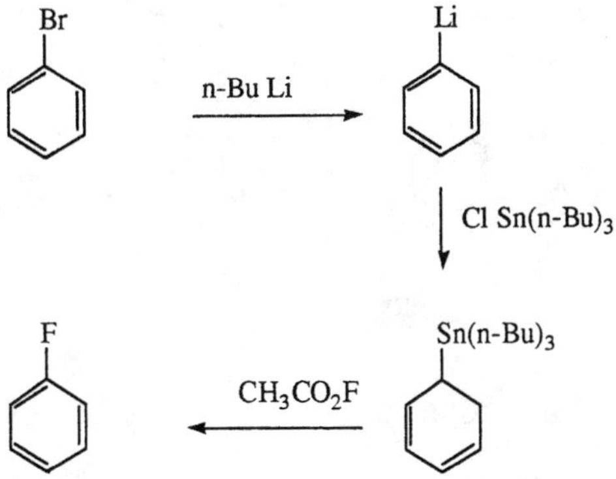

FIGURE 4.9 Synthesis and fluorination of aryl tin compounds with acetyl hypofluorite.

4.3.3 Cesium Fluoroxysulfate

Since the preparation of cesium fluoroxysulfate was first reported in 1979, considerable interest has been shown in its use as an aromatic fluorinating agent.[25] Cesium fluoroxysulfate is readily prepared by bubbling 20% F_2/N_2 through an aqueous solution of cesium sulfate at −4°C. The reagent is relatively safe to handle once prepared, although the use of metal spatulas with the dry reagent has lead to violent decomposition in some cases.

Aromatic substrates have been directly fluorinated by cesium fluoroxysulfate in the presence of a suitable catalyst. The reaction displays complex kinetics, and it is likely that it proceeds via an electrophilic attack, from which substitution or radical reactions may then follow. The Hammett value ρ+ has been determined as −3.50, compared to −2.45 for elemental fluorine. This suggests that the reagent is more selective than elemental fluorine, although the selectivity is still low.[26]

The use of boron trifluoride as a catalyst in acetonitrile has been successful for the fluorination of alkoxy-substituted benzenes and naphthalene. The relative amounts of ortho- and para-substituted products are highly dependent upon the substrate, and di-substitution is a problem in some cases.[27,28] Naphthalene derivatives are readily fluorinated, and a second fluorination reaction can lead to either further

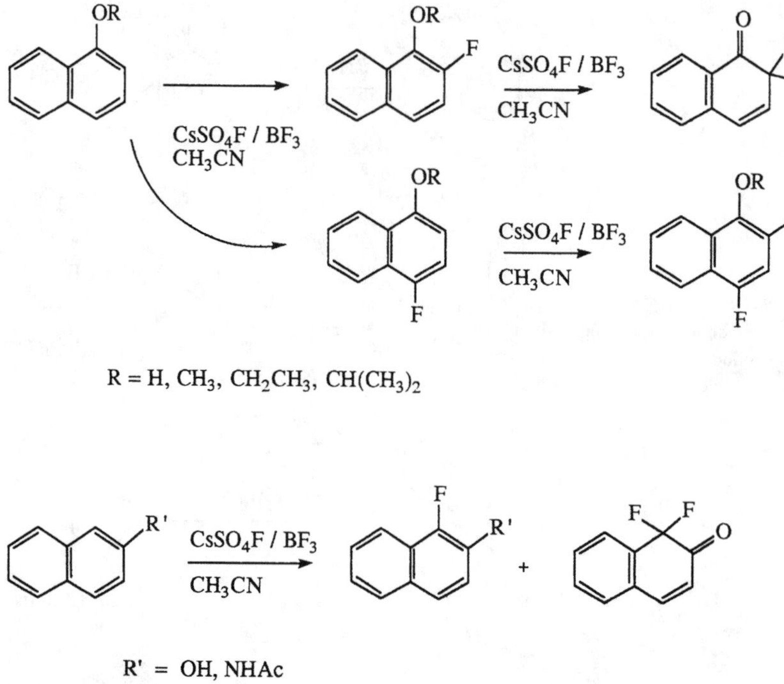

R = H, CH$_3$, CH$_2$CH$_3$, CH(CH$_3$)$_2$

R' = OH, NHAc

FIGURE 4.10 Fluorination of activated napthalenes using cesium fluoroxysulfate.

ring fluorination or loss of aromaticity, depending upon the isomer.[29] This loss of aromaticity has also been observed in the case of 2-hydroxy- and 2-*n*-acetylamino-activated napthalenes, where 1,1-difluoro-2-oxo-2,2-dihydronapthalenes are formed[30] (Figure 4.10).

Pyridine reacts with cesium fluoroxysulfate in inert solvents such as carbon tetrachloride or pentane to give a mixture of 2-fluoropyridine (70%) and 2-pyridylfluoro-sulfonate (30%). In alcoholic solvents, alkoxide-substituted pyridines, or in chlori-nated solvents (such as dichloromethane or chloroform), chloro-substituted pyridines are also produced in small amounts[31] (Figure 4.11).

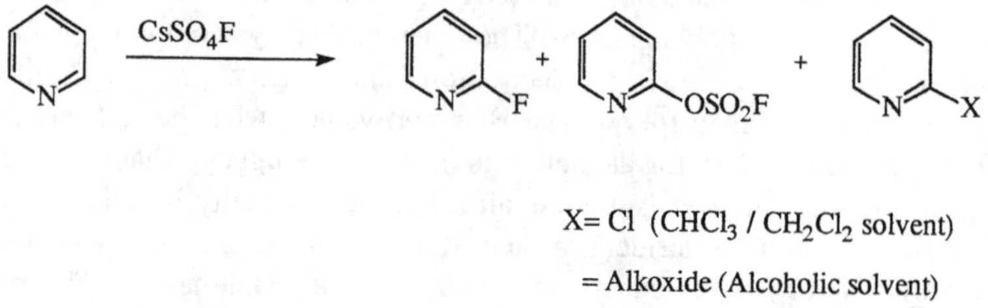

X= Cl (CHCl$_3$ / CH$_2$Cl$_2$ solvent)

= Alkoxide (Alcoholic solvent)

FIGURE 4.11 Pyridine fluorination using cesium fluoroxysulfate.

A number of other catalysts, including hydrogen fluoride, sulfuric acid, trifluoromethanesulfonic acid, fluorosulfonic acid, and antimony pentafluoride-fluorosulfonic acid have also been studied.[32] In general, the catalytic ability increases as the H° acidity parameter increases. The fluorination of toluene under catalytic conditions leads to ring fluorination, whereas the uncatalyzed fluorination leads to the formation of significant amounts of benzyl fluoride resulting from side-chain fluorination. Without catalysis, other less activated substrates often give a number of undesirable side products. α-Hydroxyalkylbenzenes, in which the ring is deactivated, are oxidized to phenones on treatment with cesium fluoroxysulfate, and alkylbenzenes containing electron-withdrawing groups are prone to side-chain fluorination. p-Alkoxy-substituted alkyl benzenes give fluoro addition products on reaction with cesium fluoroxysulfate.[33]

4.4 FLUORINATION USING XENON DIFLUORIDE

Aromatic substrates are fluorinated by xenon difluoride in the presence of catalytic amounts of hydrogen fluoride. The substitution follows the expected distributions for an electrophilic substitution reaction. Addition products such as fluorinated cyclohexadienes, cyclohexenes or cyclohexanes are not observed, although biphenyls, fluorobiphenyls, and tars may constitute significant amounts of the reaction products in some cases.[34,35]

The reaction may well proceed via a xenon difluoride-hydrogen fluoride-aromatic complex, which then generates the fluoroaromatic, or dissociates to give the radical cation which can then react with more substrate to give bi- or polyphenyls.[36] In the case of aryl oxygen compounds[37] and anilines,[38] hydrogen fluoride is not required to initiate the reaction. It is likely in this case that the aromatic ring is sufficiently activated, and has a sufficiently high oxidation potential to polarize the xenon difluoride. Electron transfer then occurs, generating the strongly oxidizing XeF˙ species, which is responsible for the fluorination (Figure 4.12). Further reaction with xenon difluoride results in formation of the fluoroaromatic and hydrogen fluoride. The side groups in aryl oxygen and aniline compounds are unaffected by the fluorination with xenon difluoride.

The reaction has been extended to heteroaromatic and polyaromatic substrates. Pyridine reacts with xenon difluoride to give a mixture of 2-fluoro (35%), 3-fluoro (20%), and 2,6-difluoropyridine (11%).[39] Polynuclear aromatics, such as naphthalene, anthracene, phenanthrene, and pyrene, also react with xenon difluoride without hydrogen fluoride initiation. Naphthalene is converted to a mixture of 1-fluoronaphthalene (50%) and 2-fluoronaphthalene (11%). Phenanthrene gives a more complex range of products, including di-, tri-, and tetrafluoro-substituted addition products.[40,41]

4.5 FLUORINATION USING HIGH-VALENCY METAL FLUORIDES

Among the high-valency metal fluorides, silver difluoride and cobalt trifluoride are probably the most important for aromatic fluorination.

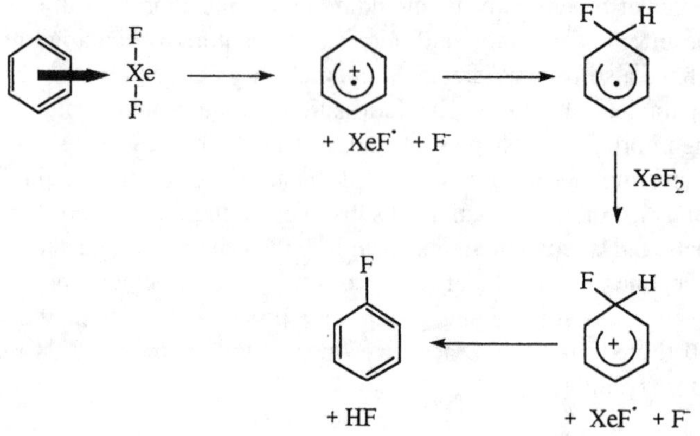

FIGURE 4.12 Mechanism of aromatic fluorination using xenon difluoride.

4.5.1 Silver (II) Fluoride

Aromatics will react with silver (II) fluoride to give a number of products.[42] The relative amounts of the various products are affected by the substrate, solvent, reaction temperature, and reaction time. Usually, the monofluorinated aromatic is the major product, although small amounts of difluoro and addition products (fluorinated cyclohexadienes) are also formed. The use of aliphatic hydrocarbon solvents such as n-hexane is preferred to halogenated solvents (chloroform, carbon tetrachloride, or dichloromethane). The reaction proceeds via oxidative addition of two fluorine atoms to give a 3,6-difluoro-1,4-dicyclohexadine, which then eliminates hydrogen fluoride to give fluorobenzene (Figure 4.13). Further addition gives rise to products such as 3,3,6,6,-tetrafluoro-1,4-cyclohexadiene.

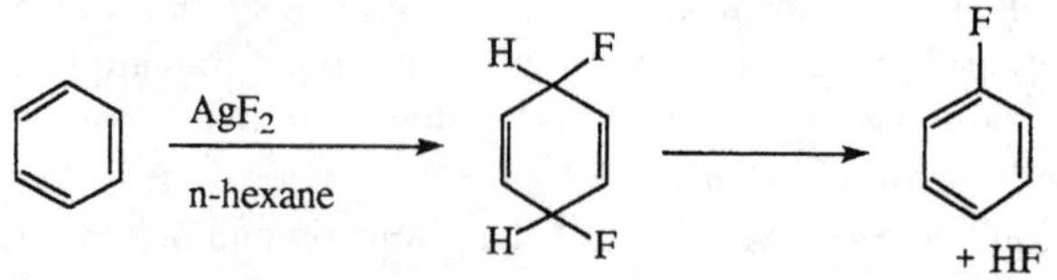

FIGURE 4.13   Aromatic fluorination using silver (II) fluoride.

4.5.2 Cobalt Trifluoride

As with electrochemical fluorination in anhydrous hydrogen fluoride (AHF), fluorination with cobalt trifluoride generally results in perfluorination. Since the heat of reaction is approximately half that of the reaction of aromatics with elemental fluorine, ring cleavage is usually not a significant side reaction. Cobalt trifluoride is generated by passing elemental fluorine over cobalt difluoride at approximately 300°C in a specially designed reactor. The substrate is then passed as a gas over the bed of cobalt trifluoride, which itself is converted back to cobalt difluoride. The bed can be reactivated by passage of more fluorine, and the cycle can be repeated a number of times before the cobalt salts require replacing[47,48] (Figure 4.14). For the synthesis of fluoroaromatics, the perfluorinated cyclohexane can be reduced by the use of a metal catalyst. Fluorination of pyridine gives undecafluoropiperidine, which can be rearomatized by passage over an iron surface at 580 to 600°C.[49] The metal (such as nickel or iron) is converted to the metal fluoride during the process. This itself may be regenerated by passing a stream of hydrogen over the surface between runs.[50]

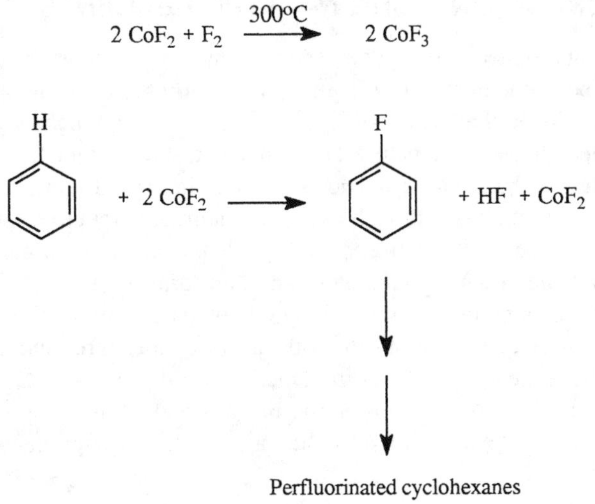

$$2 \ CoF_2 + F_2 \ \xrightarrow{300°C} \ 2 \ CoF_3$$

Perfluorinated cyclohexanes

FIGURE 4.14 Aromatic fluorination using cobalt trifluoride.

Alkali metal tetrafluoro cobalt (III) salts have also been investigated for use as aromatic fluorinating agents. They are prepared by the direct fluorination of either the cobalt trichloride (Li, Rb, Cs) or trifluoride (Na, K) salt. As with cobalt trifluoride, again the aromatic fluorination reaction gives primarily addition products, although small amounts of hexafluorobenzene are observed using cesium tetrafluoro cobalt (III).[51]

4.6 ELECTROCHEMICAL FLUORINATION

Electrochemical fluorination was originally developed by Simons during the Second World War as part of the Manhattan project. In this system, the substrate is electrolyzed in a solution of AHF.[43] The current is limited to prevent evolution of elemental fluorine. In the case of many aromatic compounds, the solution is not sufficiently conducting, and an alkali metal/alkaline earth metal electrolyte such as sodium fluoride is generally added. In general, fluorination of aromatic compounds (benzene, fluorobenzene) gives perfluorocyclohexane. Chlorobenzene is converted to chloroperfluorocyclohexane. Anisole is converted to a mixture of perfluoroethers and cleavage products.[44]

More recently, selective electrochemical fluorination has been accomplished by the use of alternative solvents. Benzotrifluoride can be converted to 3-fluorobenzotrifluoride (and some 3,5-difluorobenzotrifluoride) by electrolysis in sulfolane.[45] When AHF is not employed as the solvent, an alternative source of fluoride is required. $(C_2H_5)_4NF.3HF$ has been used in acetonitrile; the salt is electrolyzed in acetonitrile for 5 hr at 2.9 V, before addition of the benzene substrate. Further electrolysis then gives conversion to fluorobenzene and some 1,4-difluorobenzene.[46]

4.7 FLUORINATION USING N-F-TYPE REAGENTS

Of all the agents studied in recent years for organic fluorinations, the use of N-fluoro reagents has been one of the most intensive areas of research. A number of reagents have been developed that are capable of acting as electrophilic sources of fluorine (i.e., F+). Depending on the nature of the reagent, direct aromatic fluorination, or reaction with a carbanionic aromatic ring (e.g., an aryl Grignard) can give high yields of fluoroaromatics. Although elemental fluorine is required for the synthesis of the reagents, many are relatively stable once prepared, and a number are commercially available on both a laboratory and industrial scale.

Tetrafluoroammonium tetrafluoroborate is prepared from the reaction of nitrogen fluoride, fluorine, and boron trifluoride at low temperatures and ultraviolet light activation. The reagent reacts with aromatic hydrocarbons at –78°C in AHF to give a mixture of products. A range of mono-, di-, tri-, and tetrafluoroaromatics may be formed, and at high reagent-to-substrate ratios, additional products may result[52] (Figure 4.15).

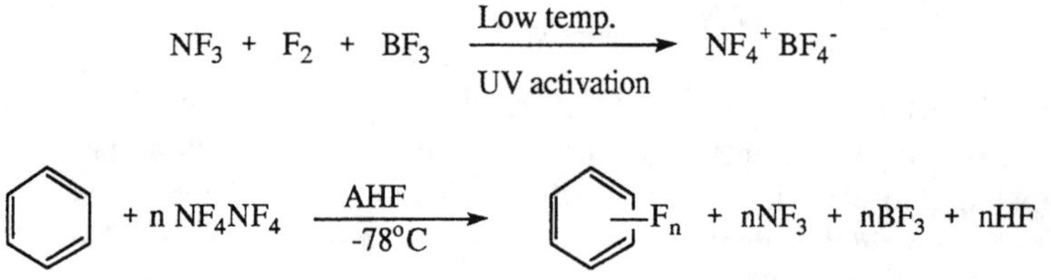

$$NF_3 + F_2 + BF_3 \xrightarrow[\text{UV activation}]{\text{Low temp.}} NF_4^+ BF_4^-$$

FIGURE 4.15   Aromatic fluorination using tetrafluoroammonium tetrafluoroborate.

N-fluoropiperidines react with sodium phenoxides to give fluorophenols. The major reaction product is actually 2,6,6-triphenoxy-1-azaperfluorocyclohexane, formed via a competitive side reaction between the fluorinating reagent and the aryl phenoxide, which limits the usefulness of the method (Figure 4.16).[53] The perfluorinated analogues, perfluoro-N-fluoropiperidines, will react with ether solutions of aryl Grignards to give fluorobenzenes, and again, significant amounts of other products result from the side reaction of the organometallic reagent with the piperidine.[54]

N-fluoroalkylsulfonamides are prepared by the reaction of alkylsulfonamides with fluorine, and have been used to fluorinate aryl Grignard reagents, and to directly fluorinate activated aromatics such as toluene[55] (Figure 4.17).

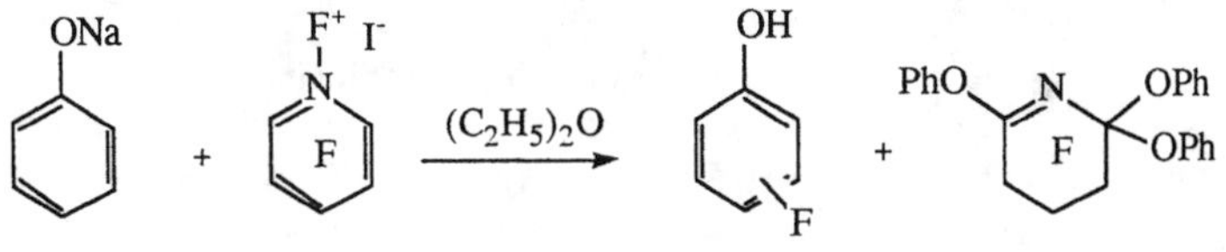

FIGURE 4.16 Fluorination of aryl phenoxides using *N*-fluoropiperidines.

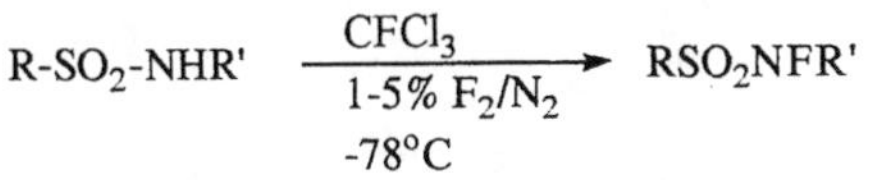

$$R\text{-}SO_2\text{-}NHR' \xrightarrow[\substack{1\text{-}5\%\ F_2/N_2 \\ -78^\circ C}]{CFCl_3} RSO_2NFR'$$

R = p-tolyl, butyl

R' = Methyl, t-butyl, exo-2-norbonyl,
     endo-2-norbonyl, cyclohexyl, neopentyl

FIGURE 4.17 Synthesis of *N*-fluoroalkylsulfonamides.

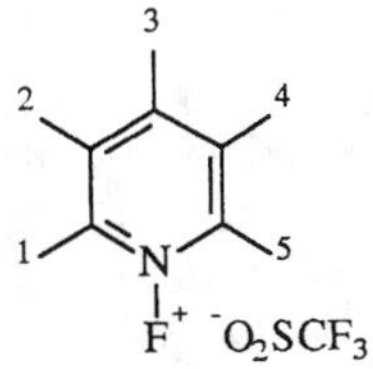

1-5	: H		
1,3,5	: CH$_3$	2,4	: H
1,3,5	: CO$_2$CH$_3$	2,4	: H
1,5	: CO$_2$CH$_3$	2,3,4	: H

FIGURE 4.18 *N*-fluoropyridinium triflates.

N-fluoropyridinium triflates are stable, nonhygroscopic, and more reactive than their tetrafluoroborate, hexafluoroantimonate, or hypochlorate analogues. The salts will rapidly fluorinate aryl Grignard reagents at room temperature, although the direct fluorination of benzene or anisole required more forcing conditions[56] (Figure 4.18). *N*-fluoro-3,5-dichloropyridinium triflate has been used to prepare biologically important fluoroaromatics from the corresponding hydrocarbon at room temperature in dichloromethane or acetonitrile[57]

A number of *N*-fluorosulfonamides have been used successfully for aromatic fluorination (Figure 4.19). *N*-fluoro-*o*-benzenedisulfonimide will convert phenyl magnesium bromide to fluorobenzene.[58] *N*-fluorobenzenesulfonimide and *N*-fluorobenzenedisulfonimide have been used to fluorinate lithiated aromatics containing a direct metalation group such as OCH$_3$, COSN(C$_2$H$_5$)$_2$, SO$_2$N(CH$_3$)$_2$, or CONH(t-C$_4$H$_9$). The aryl lithium was prepared *in situ* by treatment of the aromatic substrate with an alkyl lithium.[59] *N*-perfluoroalkylsulfonimides will convert benzene and activated aromatics to the corresponding fluoroaromatic at room temperature, with a preference for ortho substitution in some cases.[60] Of a similar reactivity is perfluoro-*N*-fluoro-(4-pyridyl)methanesulfonimide, which will also convert benzene and activated aromatics at room temperature.[61]

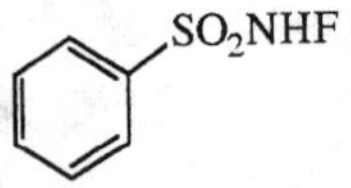

SO$_2$NHF — N-fluorobenzenesulfonimide

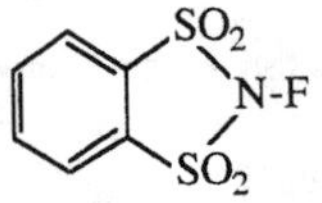

N-fluoro-o-benzenedisulfonimide

$(R_fSO_2)_2N$-F

$R_fSO_2N(F)SO_2R_f'$ N-perfluoroalkylsulfonimides

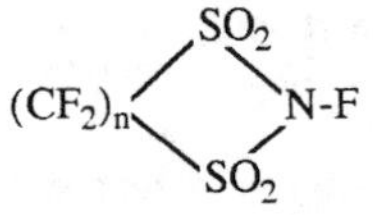

$(CF_2)_n$ ... N-F

CF$_3$SO$_2$NF

Perfluoro-N-fluoro-N-(4-pyridyl) methanesulfonimide

FIGURE 4.19   *N*-fluorosulfonamides used for aromatic fluorination.

A range of *N*-fluoropyridinium salts have been prepared, in which it is possible to control the reactivity of the fluorinating moiety. Decreasing the electron density of the nitrogen atom by modification of other ring substituents will result in an increase in the fluorinating power.[62] *N*-fluoropyridinium fluoride has also been used to fluorinate pyridines selectively in the 2-position.[63]

N-fluoroquinuclidinium fluoride, prepared by the direct fluorination of quinuclidine with fluorine at –78°C in fluorotrichloromethane (Figure 4.20), will convert aryl Grignard or aryl(trichloro)silane reagents to the corresponding aromatics, although the reagent is highly hygroscopic and conversions are low.[64,65]

A range of derivatives of *N*-fluoroquinuclidinium fluoride with varying anions (including triflate, trifluoroacetate, heptafluoro-*n*-butyrate, and tetrafluoroborate) have been synthesized to avoid problems caused by the extremely hygroscopic nature of the fluoride. The triflate and tetrafluoroborate salts are nonhygroscopic and capable of converting phenol to a mixture of 2- and 4-fluorophenol, as well as the fluorination of aryl Grignards previously discussed.[66,67]

1-Alkyl-4-fluoro-1,4-diazoniabicyclo[2.2.2]octane salts[68] (Figure 4.21) are commercially available under the tradename Selectfluor™.[69] These nonhygroscopic solids are soluble in a number of polar solvents, including water, acetonitrile, and *N,N*-dimethylformamide, and are specifically marketed for selective, electrophilic substitution under mild conditions.

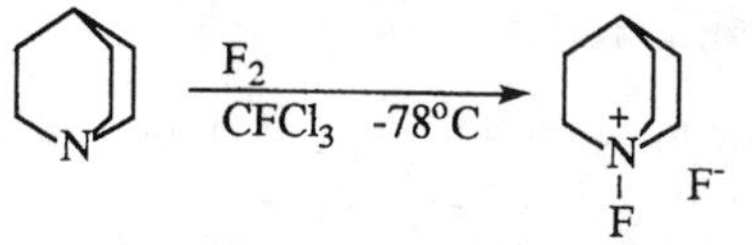

FIGURE 4.20   *N*-fluoroquinuclidium fluoride.

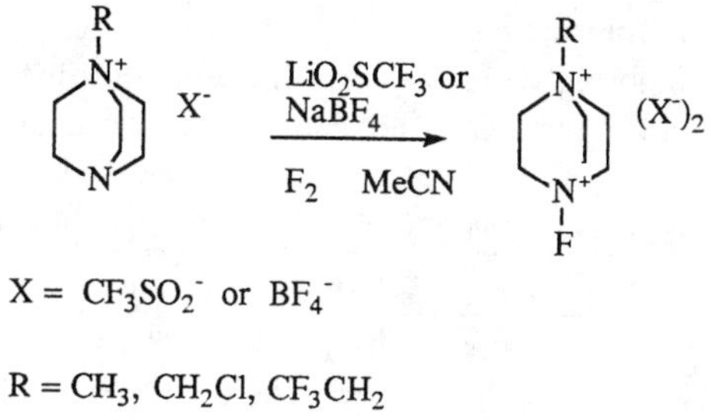

FIGURE 4.21   Synthesis of 1-alkyl-4-fluoro-1,4-diazoniabicyclo[2.2.2] octane salts (Select-fluor™).

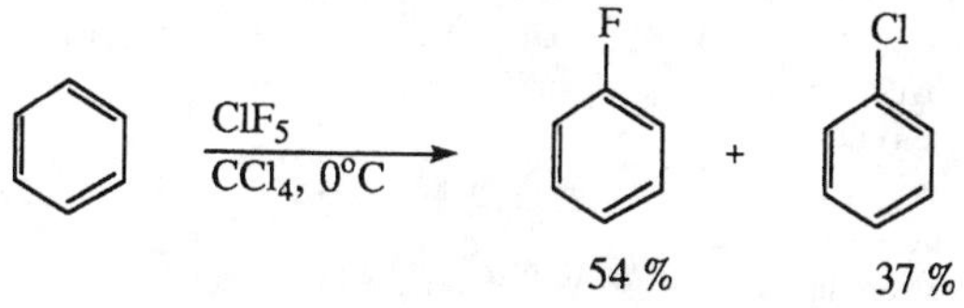

$X = CF_3SO_2^-$ or BF_4^-

$R = CH_3, CH_2Cl, CF_3CH_2$

FIGURE 4.22   Aromatic fluorination using chlorine pentafluoride.

54 %   37 %

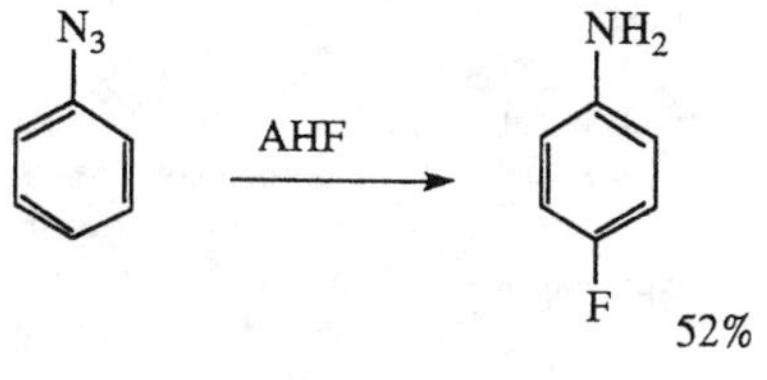

FIGURE 4.23   Aromatic fluorination in anhydrous hydrogen fluoride activated by azide groups.

52%

4.8   MISCELLANEOUS FLUORINATION METHODS

Chlorine pentafluoride has been used to directly fluorinate aromatics (including heterocyclics), although chlorination is often a significant side reaction accompanying fluorination[70] (Figure 4.22). Chlorine trifluoride has also been used, although addition coupling and polymeric products are also obtained.[71]

Fluorinated fullerenes ($C_{60}F_{44}$ to $C_{60}F_{46}$) will react with aromatics in the presence of a Lewis acid catalyst (boron trifluoride etherate) to give fluoroaromatics and hydrogen fluoride, albeit in low yields. They will, however, convert aryl lithium compounds to the corresponding fluoroaromatic in reasonable yields.[72]

Azides, prepared from the corresponding diazonium by treatment with sodium azide, will activate aromatic rings towards fluorination in anhydrous hydrogen fluoride, to give the *para*-fluoroaniline[73] (Figure 4.23).

4.9 SYNTHETIC METHODS

The major difficulty encountered with electrophilic fluorination is in the use of elemental fluorine. The element is highly toxic and corrosive, meaning that special handling conditions, equipment, and protection for those handling it are essential. This does limit the use of fluorine to those experienced in its handling, precluding use by many mainstream organic chemists. Since the particular effects fluorine imparts on a molecule are attractive to many areas of chemistry, the development of fluorinating agents that can be more readily handled is essential. The most important reagents used for aromatic fluorination are listed in Table 4.1, together with a number of advantages and disadvantages.

Table 4.1 Synthetic Methods for Aromatic Fluorination

Reagent	Advantages	Disadvantages
Elemental fluorine	Inexpensive, readily available; will fluorinate wide range of aromatics	Toxic, corrosive, requires special handling conditions; reactions require moderating (low temperature, solvent, diluted F_2)
Trifluoromethyl hypofluorite/acetyl hypofluorite (more selective)	React rapidly with organometallic/organosilanes; useful for ^{18}F radiolabeling	Expensive, toxic, corrosive, explosive
Cesium fluoroxysulfate	More selective than F_2	Commercially unavailable; requires F_2 for synthesis; often requires catalyst
Xenon difluoride	Will fluorinate highly activated aromatics directly	Very expensive; often requires AHF catalyst
High-valency metal fluorides	Useful for perfluorination	Need to re-aromatize after fluorination; unsuitable for selective fluorination
Electrochemical fluorination	Useful for perfluorination with AHF; use of fluoride salts can lead to selective fluorination	Need to re-aromatize after fluorination; requires special electrolysis equipment
N-F-type reagents	Can be obtained commercially; avoids use of elemental F_2; often very selective	May only fluorinate organometallics/activated aromatics

Although elemental fluorine is difficult to handle, it is obviously much cheaper than reagents that must be prepared from the element. On a commercial scale, this makes direct fluorination the most obvious method. However, a number of additional problems arise from the use of fluorine. Since the element is so reactive, reactions must generally be moderated in some way. In some cases, explosive reactions may result. In general, dilution with an inert solvent, such as nitrogen or argon, is used, often so that the amount of fluorine in the stream is between 2 and 20% (v/v). The inert gas also acts as a thermal sink for the exothermic reaction. The reaction is generally carried out below room temperature, and sometimes as low as −78°C, which adds more practical difficulties, since efficient cooling is essential. Generally, a solvent is required, which itself will be unaffected by fluorine. Solvents successfully used include acetonitrile, methanol, and fluorocarbon solvents such as hexafluorobenzene, chloroform, and freons. More recently, the use of protic acids has been successful.[8] Direct aromatic fluorination proceeds via an electrophilic substitution mechanism, leading to ortho- and para-substituted products in the case of substituted aromatics. In the case of a multistep synthesis, in which fluorination occurs close to (or at) the final step, regioselective fluorination is essential. However, aryl oxygen-substituted aromatics do give strong ortho selectivity under direct fluorination, and para regioslectivity may result from the addition of a Lewis acid catalyst.

The replacement of other groups at carbanionic carbon centers results in more selective fluorination. However, the use of this approach has two major drawbacks. The first is that the organometallic or silyl group used often must be selectively introduced into the ring, creating additional steps in the reaction sequence. The second is the disposal of the waste metal (such as tin or thallium) salts, which, while practical on laboratory scale, may be more expensive than simply separating and disposing of the unwanted isomer on a commercial scale. The particular use of this route is for ^{18}F-radiolabeled compounds, where the dominant factors are the reaction time and radiochemical yield. Since the half-life of ^{18}F is only 120 min, fluorination must be as rapid as possible, which can be readily achieved using elemental fluorine. Since the displacement reaction is regiospecific, time-consuming work-up and purification processes are also minimized. The Balz-Schiemann reaction is also used for the synthesis of radiolabeled fluoroaromatics; however, since only a quarter of the ^{18}F-labeled BF_4 actually ends up in the aromatic product, the use of ^{18}F-labeled F_2 (where half the ^{18}F is incorporated) gives twice the radiochemical yield.

Reagents such as trifluoromethyl hypofluorite, bis(fluoroxy)trifluoromethane, and the more selective acetylhypofluorite, together with cesium fluoroxysulfate, are limited by their commercial availablilty, their cost, and their high toxicity and, in some cases, may be explosive. Radiolabeled acetylhypofluorite has been used successfully to gave rapid conversion to labeled fluoroaromatics, in some cases more selectively than with fluorine. However, the synthesis uses radiolabeled elemental fluorine and requires additional time (and corresponding loss of radioactivity). Dry cesium fluoroxysulfate has been reported to explode in contact with metallic spatulas, although otherwise it is relatively straightforward to handle. The use of cesium fluoroxysulfate for aromatic fluorination generally requires a Lewis acid catalyst such as boron trifluoride, although pyridines may be fluorinated without catalysis.

Although xenon difluoride is a useful fluorinating agent, it is very expensive, and in some cases, anhydrous hydrogen fluoride (AHF) is required as an initiator. AHF is highly corrosive and toxic; special precautions for its handling and use are required, as with elemental fluorine. This limits the applicability of the method for use by nonspecialists. Although additional products are not observed, tars may often be formed in significant quantities.

High-valency metal fluorides were used during the 1950s and 1960s by industry for the commercial synthesis of fluoroaromatics, including hexafluorobenzene and 1,2,3,4-tetrafluorobenzene. The synthesis of perfluoroaromatics is now carried out on a commercial scale via halex methodology. The reaction is useful for continuous operation, since substrate and fluorine (to regenerate the reagent) can be passed alternatively over a bed to the metal fluoride. Since the reactions generally give perfluorination, and loss of aromaticity, re-aromatization via passage over a metal catalyst is required. This means that the method has little application for selective fluorination.

Electrochemical fluorination in AHF again gives perfluorination, and limits the applicability of the process for selective fluorination. AHF, like elemental fluorine, is inexpensive when compared to other reagents such as xenon difluoride or trifluoromethyl hypofluorite. Electrochemical fluorination using other salts, such as tetraethylammonium fluoride-hydrogen fluoride complexes, gives selective fluorination, although the process still requires specialist equipment for the electrolysis, and is not particularly suitable for small-scale synthetic work.

Perhaps the most useful reagents to emerge in recent years have been the N-F class. Although the synthesis of these compounds requires elemental fluorine, many are relatively safe to handle once prepared. A number of electrophilic reagents are commercially available, on scales ranging from the laboratory to the industrial level. These offer the first real capability for electrophilic fluorinations that can be conducted in standard laboratory glassware, without the necessity for very low temperatures or high dilutions. As with many of the other reagents available, they fluorinate in accordance with electrophilic substitution patterns, which may lead to mixtures of products. Milder reagents in this class have been used to replace organometallic groups, such as Grignards, which are more readily prepared (or commercially available) and less toxic than the organotin and organothallium compounds used with other fluorinating systems.

Table 4.2   Aromatic Fluorinations Using Selectfluor™

Substrate	Reaction Conditions	Product	Yield/Selectivity
NHCOCH$_3$	CH$_3$CN reflux, 5 min	NHCOCH$_3$ —F	80% yield o:p = 62:38
CH$_3$	CH$_3$CN reflux, 16 hr	CH$_3$ —F	82% yield o:p = 3:1
OCH$_3$	CH$_3$CN reflux, 3 hr	OCH$_3$ —F	96% yield o:p:difluoro = 2:1:1
OH	CH$_3$CN room temp., 15 hr	OH —F	24% yield o:p = 2:1

Selectfluor™ reagent F-TEDA-BF$_4$ (1-chloromethyl-4-fluoro-1,4-diazoniabi-cyclo[2.2.2] ocatane bis(tetrafluoroborate) (Air Products) is marketed specifically for selective fluorination, particularly in pharmaceutical and agricultural applications, and is a white, free-flowing, nonhygroscopic powder that can be handled with ease. Examples of aromatic fluorination using F-TEDA-BF$_4$ are given in Table 4.2.[69] Other reagents available commercially include N-fluoroquinuclidium fluoride, N-fluoro-N-alkyl-p-toluenesulfonamides, and N-fluoropyridinium salts such as N-fluoropyridinium pyridine heptafluorodiborate (Allied Signal).[67]

REFERENCES

1. A.H. Vasek and L.C. Sams, *J. Fluorine Chem.*, 1972, 2, 257.
2. A.H. Vasek and L.C. Sams, *J. Fluorine Chem.*, 1973/4, 3, 397.
3. C.L. Jeffrey and L.C. Sams, *J. Org. Chem.*, 1977, 42, 863.
4. F. Cacace and A.P. Wolf, *J. Am. Chem. Soc.*, 1978, 100, 3639.
5. V. Grakauskas, *J. Org. Chem.*, 1970, 35, 723.
6. S. Misaki, *J. Fluorine Chem.*, 1981, 17, 159.
7. S. Misaki, *J. Fluorine Chem.*, 1982, 21, 191.
8. R.D. Chambers, C.J. Skinner, J. Thompson and J. Hutchinson, *J. Chem. Soc. Chem. Commun.*, 1995, 17.
9. M. VanDerPuy, *Tetrahedron Lett.*, 1987, 28, 235.
10. S.T. Purrington and D.L. Woodward, *J. Org. Chem.*, 1991, 56, 143.
11. L.C. Sams, T.A. Reames and M.A. Durrance, *J. Org. Chem.*, 1978, 43, 2273.
12. P. DiRaddo, M. Diksic and D. Jolly, *J. Chem. Soc. Chem. Commun.*, 1984, 159.
13. M. Speranza, C.Y. Shuie, A.P. Wolf, D.S. Wilbur and G. Angelini, *J. Fluorine Chem.*, 1985, 30, 97.
14. J. DeYoung, H. Kawa and R.J. Lagow, *J. Chem. Soc. Chem. Commun.*, 1992, 811.
15. M.J. Adam, B.D. Pate, T.J. Ruth, J.M. Berry and L.D. Hall, *J. Chem. Soc., Chem. Commun.*, 1981, 733.
16. M.J. Adam, J.M. Berry, L.D. Hall, B.D. Pate and T.J. Ruth, *Can. J. Chem.*, 1983, 61, 658.
17. M.K. Fifoly, R.T. Olczak and R.F. Mundhenke, *J. Org. Chem.*, 1985, 50, 4576.
18. M.R. Bryce, R.D. Chambers, S.T. Mullins and A. Parkin, *J. Fluorine Chem.*, 1984, 26, 533.
19. O. Lerman, Y. Torr and S. Rozen, *J. Org. Chem.*, 1981, 46, 4629.
20. O. Lerman, Y. Torr, D. Hebel and S. Rozen, *J. Org. Chem.*, 1984, 49, 806.
21. M. Speranza, C.Y. Shiue, A.P. Wolf, D.S. Wilbur and G. Angelini, *J. Chem. Soc. Chem. Commun.*, 1984, 1448.
22. G.W.M. Visser, B.W.V. Halteren, J.D.M. Hercheid, G.A. Brinkman and A. Hoekstra, *J. Chem. Soc. Chem. Commun.*, 1984, 665.
23. G.W.M. Visser, C.N.M. Bakker, B.W.V. Halteren, J.D.M. Hersceid, G.A. Brinkman and A. Hoekstra, *J. Org. Chem.*, 1986, 51, 1886.
24. M.J. Adam, T.J. Ruth, S. Jivan and B.D. Pate, *J. Fluorine Chem.*, 1984, 25, 329.
25. E.H. Appleman, L.J. Basile and R.C. Thompson, *J. Am. Chem. Soc.*, 1979, 101, 3384.
26. D.P. Ip, C.D. Arthur, R.E. Winans and E.H. Appelman, *J. Am. Chem. Soc.*, 1981, 103, 1964.
27. S. Stavber and M. Zupan, *J. Chem. Soc. Chem. Commun.*, 1981, 148.
28. S. Stavber and M. Zupan, *J. Fluorine Chem.*, 1981, 17, 597.
29. J. Stavber and M. Zupan, *J. Org. Chem.*, 1985, 50, 3609.
30. T.B. Patrick and D.L. Darling, *J. Org. Chem.*, 1986, 51, 3242.
31. S. Stavber and M. Zupan, *Tetrahedron Lett.*, 1990, 31, 775.
32. E.H. Appleman, L.J. Basile and R. Hayatsu, *Tetrahedron*, 1984, 40, 1892.
33. S. Stavber, Z. Planinsek, I. Kosir and M. Zupan, *J. Fluorine Chem.*, 1992, 59, 409.
34. M.J. Shaw, H.H. Hyman and R. Filler, *J. Am. Chem. Soc.*, 1969, 91, 1563.
35. M.J. Shaw, H.H. Hyman and R. Filler, *J. Am. Chem. Soc.*, 1970, 92, 6498.
36. M.J. Shaw, H.H. Hyman and R. Filler, *J. Org. Chem.*, 1971, 36, 2917.

37. S.P. Anand, L.A. Quarterman, H.H. Hyman, K.G. Migliorese and R. Filler, *J. Org. Chem.*, 1975, 40, 807.
38. S.P. Anand and R. Filler, *J. Fluorine Chem.*, 1976, 7, 184.
39. S.P. Anand and R. Filler, *J. Fluorine Chem.*, 1976, 7, 184.
40. S.P. Anand, L.A. Quarterman, P.A. Christian, H.H. Hyman and R. Filler, *J. Org. Chem.*, 1975, 40, 3796.
41. E.D. Bergmann, H. Selig, C.H. Lin, M. Rabinovitz and I. Agranat, *J. Org. Chem.*, 1975, 40, 3793.
42. A. Zweig, R.G. Fischer and J.E. Lancaster, *J. Org. Chem.*, 1980, 45, 3597.
43. J. Burdon and J.C. Tatlow, *Adv. Fluorine Chem.*, 1960, 1, 129.
44. Y. Inoue, S. Nagase, K. Kodaira, H. Baba and T. Abe, *Bull. Chem. Soc. Jpn.*, 1973, 46, 2204.
45. V.A. Shreider and I.N. Rozhkov, *Izv. Akad. Nauk. SSSR. Ser. Khim.*, 1976, 676.
46. I.N. Rochkov, A.V. Bukhtiarov and I.L. Knunyants, *Izv. Akad. Nauk SSSR Ser. Khim.*, 1972, 1130.
47. M. Stacey and J.C. Tatlow, *Adv. Fluorine Chem.*, 1960, 1, 166.
48. R.N. Haszeldine and F. Smith, *J. Chem. Soc.*, 1950, 3617.
49. R.E. Banks, A.E. Ginsberg and R.N. Haszeldine, *J. Chem. Soc.*, 1961, 1740.
50. B. Gething, C.R. Patrick, M. Stacey and J.C. Tatlow, *Nature*, 1959, 183, 588.
51. A.J. Edwards, R.G. Plevey, I.J. Sallomi and J.C. Tatlow, *J. Chem. Soc. Chem. Commun.*, 1972, 1028.
52. C.J. Schack and K.O. Christe, *J. Fluorine Chem.*, 1981, 18, 363.
53. V.R. Polishchuk and L.S. German, *Tetrahedron Lett.*, 1972, 51, 569.
54. R.E. Banks, V. Murtagh and E. Tsiliopoulos, *J. Fluorine Chem.*, 1991, 52, 389.
55. W.E. Barnette, *J. Am. Chem. Soc.*, 1984, 106, 452.
56. T. Umemoto, K. Kawada and K. Tomita, *Tetrahedron Lett.*, 1986, 27, 4465.
57. D. Hebel and K.L. Kirk, *J. Fluorine Chem.*, 1990, 47, 179.
58. F.A. Davis and W. Han, *Tetrahedron Lett.*, 1991, 32, 1631.
59. V. Snieckus, F. Beaulieu, K. Mohri, W. Han, C.K. Murphy and F.A. Davis, *Tetrahedron Lett.*, 1994, 35, 3465.
60. S. Singh, D.D. Desmarteau, S.S. Zuben, M. Witz and H. Huang, *J. Am. Chem. Soc.*, 1987, 109, 7194.
61. R.E. Banks and A. Khazali, *J. Fluorine Chem.*, 1990, 46, 297.
62. T. Umemoto, S. Fukami, G. Tomizawa, K. Harasawa, K. Kawada and K. Tomita, *J. Am. Chem. Soc.*, 1990, 112, 8563.
63. A.S. Kiselyov and L. Strekowski, *J. Org. Chem.*, 1993, 58, 4476.
64. R.E. Banks, R.A. DuBoisson and E. Tsiliopoulos, *J. Fluorine Chem.*, 1986, 32, 461.
65. R.E. Banks, R.A. DuBoisson, W.D. Morton and E. Tsiliopoulos, *J. Chem. Soc. Perkin Trans. 1*, 1988, 2805.
66. R.E. Banks and I. Sharif, *J. Fluorine Chem.*, 1988, 41, 297.
67. R.E. Banks and I. Sharif, *J. Fluorine Chem.*, 1991, 55, 207.
68. R.E. Banks, S.N. Mohialkin Khaffaf, F. Lal, I. Sharif and R.G. Syvret, *J. Chem. Soc.* Chem. Commun., 1992, 595.
69. Selectfluor™ trademark of AirProducts and Chemicals Inc., Allentown, PA, USA.
70. M.M. Boudakian and G.A. Hyde, *J. Fluorine Chem.*, 1984, 25, 435.
71. W.K.R. Musgrave, *Adv. Fluorine Chem.*, 1960, 1, 21.
72. A.A. Gakg, A.A. Tuinman, J.L. Adcock and R.N. Compton, *Tetrahedron Lett.*, 1993, 34, 7267.
73. D.M. Mulvey, A.M. Demarco and L.M. Weinstock, *Tetrahedron Lett.*, 1978, 16, 1419.

Chapter 5

Trifluoromethylaromatics

5.1 INTRODUCTION

The trifluoromethyl group is now considered to be second in importance only to fluorine itself as a fluorine-containing aromatic substituent. Inspection of both the primary scientific and patent literatures confirms this, and reveals a steady increase in related work over the last 10 years.[1,2] Trifluoromethylated aromatic compounds are the subject of considerable interest in several successful industrial sectors, including pharmaceuticals, agrochemicals, dyes, liquid crystals, and polymers. In some of these (e.g., agrochemicals), there are now established trifluoromethylated aromatic products in the market, while in others (e.g., polymers), new products are largely at the research stage.

The most commonly quoted effect of the trifluoromethyl group on biologically active molecules is increased lipophilicity leading to enhanced solubility in fatty tissue and more efficient transport in the body. The high electronegativity of the group and its relatively small size (only two and a half times the volume of the methyl group) are also important factors but the added value of high molecular lipophilicity in particular gives greater benefit when compared to fluorine itself. Trifluoromethylation is known to increase light fastness as well as a shift in color (both bathochromic and hypochromic), providing obvious product value in the dye industry. The high electron-withdrawing ability and small size of the trifluoromethyl group are key factors in making compounds such as the trifluoromethyl-substituted alkylbiphenyls, useful as liquid crystals. In the context of polymers, high chemical and thermal stability are fundamentally important properties that can be enhanced by the selective introduction of trifluoromethyl groups into aromatic moieties. But it is the special effects that trifluoromethylation can have on the water resistance (reducing water uptake) and the electrical properties (reducing electric dissipation) of materials that are likely to lead to future commercial exploitation in niche areas. It is commercially significant that, even at very low levels of fluorine content (e.g., in copolymers), useful effects on material properties can be achieved.

Concomitant with the increased level of application-oriented research into trifluoromethylated aromatics, there has been renewed vigor in the design of new trifluoromethylation methods that offer improvements over traditional methods in terms of cost, simplicity, efficiency, versatility, or environmental aspects through the use of more benign reagents and/or the generation of less waste. Synthetic routes to trifluoromethylated aromatic molecules can be largely divided into several areas:

- Halogen exchange (traditional)
- Organometallic (source of CF_3^- most famously involving $CuCF_3$)
- Radical
- Other CF_3 sources (e.g., electrophilic trifluoromethylating agents)
- Chemical conversion of a simple CF_3-containing building block

The halogen exchange method, which starts from a methyl group and goes via the trichloromethyl intermediate to the final product, represents one of the oldest areas of fluorine technology and remains an important industrial method. Of the other methods, the use of $CuCF_3$ (which is unstable and needs to be generated *in situ*) is perhaps the most likely to be used on a significant level in the future, although there is still considerable scope for the development of new methods. The range of methods is sufficiently broad and diverse to enable the synthesis of trifluoromethylated aromatics from the most common starting materials, i.e.,

$$ArX \rightarrow ArCF_3$$

$(X = CH_3, \text{halogen}, CO_2H, H)$

5.2 PHYSICAL PROPERTIES

The trifluoromethyl group is larger than the methyl group but significantly smaller than the trichloromethyl group (Table 5.1).[3] Perhaps surprisingly the replacement of a CH_3 group by CF_3 has little or no effect on the bond length between the group and the aromatic nucleus. The low steric impact of CF_3 is particularly important in terms of pharmacological properties, since the replacement of CH_3 by CF_3 should result in minimal disruption to an enzyme-substrate complex.

Table 5.1   Sizes of Methyl Groups

	CH_3	CF_3	CCl_3
Van der Waals radius/Å	2.0	2.7	3.5
Van der Waals volume (hemisphere)/Å³	16.8	42.5	91.2

Equally important to its minimal steric demands is the low hydrophobicity that the trifluoromethyl group imparts on aromatic molecules. This can be measured via the relative partition coefficients of variously meta-substituted 3-phenoxyacetic acids between water and octan-1-ol (Table 5.2). Low hydrophobicity is commonly linked to high lipophilicity and an enhanced ability to transverse lipid membranes. This will lead to more rapid and efficient transport of drugs to the target receptors. Values must be treated with some caution, however, as lipophilicity is dependent upon the position of the substituent in the molecule.[4]

High electronegativity, leading to changes in molecular electronic distributions and molecular reactivity, is commonly associated with fluorine and fluorine-containing groups, and the trifluoromethyl group is no exception. While not as high as fluorine itself, it is comparable to oxygen and greater than chlorine (Table 5.3). The attractiveness and utility of CF_3 as a substituent in a biologically active molecule

Table 5.2 Hydrophobic Parameters of Aromatic Substituents (Based on 1-octanol-Water Partition Coefficients)

H	1.8
F	2.3
CH_3	2.5
CF_3	2.8
Cl	2.8
CCl_3	2.9
SCF_3	3.8

are in part a result of profound effects on molecular electronic properties, as well as the relatively small steric demands, and the enhancement in molecular lipophilicity. Fluorine itself is certainly capable of satisfying the first two of these effects, but is less effective in its rather small effects on molecular transport *in vivo*.

Table 5.3 Pauling Electronegativities for Aromatic Substituents

H	2.1
CH_3	2.3
OCH_3	2.7
CF_3CH_2	2.9
Cl	3.0
O	3.5
CF_3	3.5
OCF_3	3.7
F	4.0

Electronic factors are also of prime importance in the design of liquid crystal materials. The small relative dielectric anisotropy required in ferroelectric liquid crystal displays is achieved by the introduction of a lateral polar substituent.[5,6] This is commonly achieved by the use of a CN group but the CF_3 group is an alternative. Spontaneous polarization, a measure of the transverse dipole moments in ferroelectric materials, can also be increased by the use of CF_3, as can birefringence (a measure of the difference between the refractive index parallel to the molecular axis and the refractive index perpendicular to the molecular axis).[7]

While the subject is still at a very early stage in its development, it may well be that the effects of fluorine on the physical properties of polymeric materials will have as profound an effect on the advanced materials industry as fluorine is now having on the pharmaceutical industry. The trifluoromethyl group is already proving to be especially important in this context.

Electronic effects are likely to be of less direct importance to polymer properties than to the properties of liquid crystals, although if the fluorine-containing group, such as CF_3, is present in the monomer, this is likely to influence monomer reactivity in the actual polymerization process. Thus, base-catalyzed condensation polymerizations leading to polyaromatics such as polyetherketones can either benefit or lose by the presence of ring-trifluoromethyl groups (Figure 5.1). The reactivity of the electrophile (e.g., a 4,4′-dihalobiphenyl) will be significantly enhanced by CF_3. This could be useful so as to reduce the high temperatures normally required for such reactions (>200°C). The presence of CF_3 groups on the nucleophile could, however, be detrimental and might even prevent the process from being viable. This oversimplifies the matter as the solubility of the polymer is often the most influential parameter in determining the temperature at which the polymerization process is run (e.g., use of a high temperature to prevent premature precipitation of a relatively insoluble polymer). Of course, the CF_3 group will also affect solubility of both the monomer and the polymer (usually for the better)!

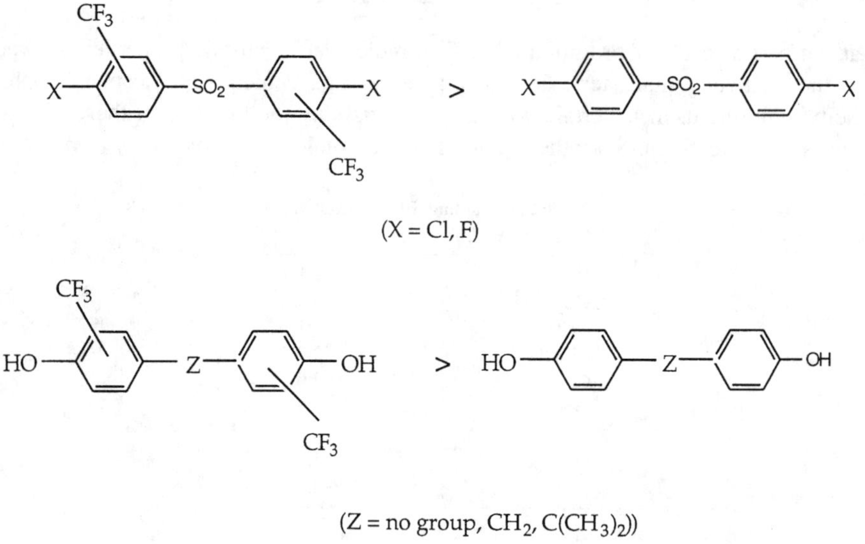

(X = Cl, F)

(Z = no group, CH_2, $C(CH_3)_2$))

FIGURE 5.1 Effects of trifluoromethylation on monomer reactivity.

Much of the original research on selectively fluorinated aromatic polymers involved the use of the hexafluoroisopropylidene group, $-C(CF_3)_2-$,[8,9] largely a result of the ease of synthesis of an aromatic-$C(CF_3)_2$-aromatic unit by reaction of suitable aromatics such as phenol and aniline with hexafluoroacetone, $(CF_3)_2C = O$. Various novel polyaromatics, and polyimides in particular, involving this bridge have been prepared and found to have useful properties such as excellent solubility and low dielectric constant.[10,11] More recently, however, the area has expanded to include trifluoromethylated aromatic units within selectively fluorinated polyaromatics. These are normally synthesized via preformed trifluoromethylated aromatics. Trifluoromethylation of a polymer is an alternative method that has yet to be properly explored. Polyimide materials in this category have again attracted considerable attention and the often outstanding effects that the CF_3 group can have on the physical properties of the polymers are especially striking. Some of the most important and interesting of these effects are summarized in Table 5.4. The loss in thermal stability of polymers on trifluoromethylation is a drawback, but this is not always significant and improvements in hydrophobicity, solubility, and other key polymer properties will usually outweigh this small disadvantage.

Table 5.4   Important Changes in Physical Properties of Polyaromatics Containing Trifluoromethyl Group

Property	Examples
Reduced water absorption/easier drying	Reduction in water uptake by factor of 3 in copolyimides[13]
	Reduction in water uptake by factor of 2 and very easy drying[14]
Excellent polymer solubility	Completely soluble copolyimides in amide solvents[13]
	Poly(aryl ether oxazole)s soluble in various organic solvents
Good miscibility with other polymers	CF_3-polyamides miscible with various nylons[15]
Reduced thermal stability	Stability of poly(aryl ether oxadole)s reduced by >100°C on trifluoromethylation[16]
Low dielectric constant	CF_3-polyimides have very low dielectric constants[12,17]
High optical transparency	Refractive indices can be controlled by adjusting CF_3 content in polyimides[12,17]
Variable coefficient of thermal expansion	Controllable CTE by adjusting CF_3 content in polyimides[12,17]

Trifluoromethylated and related selectively fluorinated polyaromatics (it is only a matter of time before other small fluorine-containing groups are used as substituents in polyaromatics) is a rapidly expanding area with potential applications for products in many specialist areas that can tolerate the high cost of the materials, including interlayer dielectrics in microelectric devices,[12] insulative coatings for electronic packaging,[13] advanced materials, adhesives, and others.[1,2]

5.3 CHEMICAL PROPERTIES

The incorporation of the highly electron-withdrawing trifluoromethyl group into an aromatic molecule can markedly affect the chemistry of that molecule. As in aliphatic chemistry, good illustrations of this can be seen in Brönsted acidities. Thus, trifluoromethylated phenols are clearly stronger Brönsted acids, but also weaker Brönsted bases, than phenol itself. Extreme examples are known. Thus, pentakis(trifluoromethyl)cyclopentadiene is a remarkably strong Brönsted acid (pK_a <–2) comparable to nitric acid (Figure 5.2). Trifluoromethyl groups can be especially valuable in this context, since a fluorine directly bonded to a carbanion center is destabilizing due to electron-electron repulsions between the negative charge and the fluorine lone pairs.

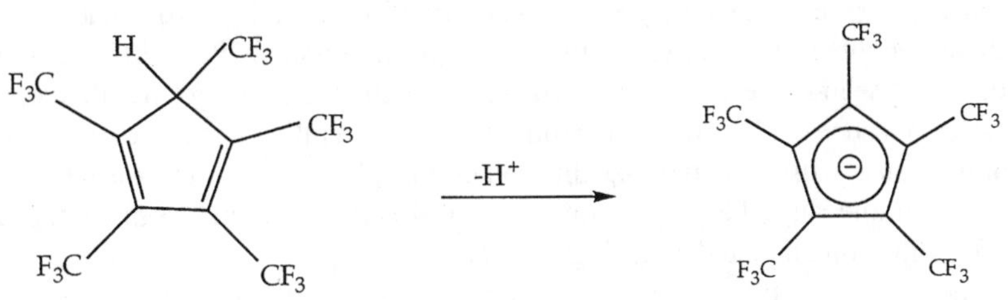

FIGURE 5.2   Pentakis-(trifluoromethyl)cyclopentadiene.

Trifluoromethyl groups will also activate other ring substituents towards nucleophilic attack. Thus, while nitrobenzene itself is resistant to nucleophilic substitution of the nitro group by fluoride, 2-nitrobenzotrifluoride can undergo nucleophilic fluorodenitration under powerful fluoride ion conditions (this may also be a result of the CF_3 group twisting the nitro group out of the plane of the ring, making it a better leaving group — see Chapter 2).[18] It should be noted, however, that the CF_3 group exerts only a –I effect, so that halogen exchange reactions and other nucleophilic substitutions on halogenobenzotrifluorides will require more forcing conditions than the corresponding halonitrobenzenes. Fluorobenzotrifluorides are more reactive than the corresponding chlorobenzotrifluorides.

Replacement of chlorine at the 4-position in 4-chloro, 4-chloro-3-nitro-, and 3,4-dichlorobenzotrifluorides using O-centered nucleophiles such as phenols is an important reaction for the preparation of herbicides.

The trifluoromethyl group strongly deactivates aromatic substrates towards electrophilic attack and is strongly meta directing. Benzotrifluoride is 10^2 to 10^5 less reactive than benzene (nitrobenzene is 10^3 to 10^6 less reactive). The meta-directing effect can be overcome by a sufficiently powerful ortho/para-directing group. In fact, small but significant amounts of ortho-substitution occur in electrophilic substitution reactions with benzotrifluorides, which can cause problems with large-scale reactions, especially in these days of "waste minimization".

Nitration of benzotrifluoride with the usual mixture of concentrated nitric and sulfuric acids at 0°C gives about 90% of the *meta*-nitrobenzotrifluoride. The resulting isomer mixture is rather difficult to separate and it is usually better to reduce the mixture to the corresponding anilines before separation is undertaken (anilines may be easier to separate because of their hydrogen bonding activity). Dinitration of 4-chlorobenzotrifluoride, to give the important intermediate 4-chloro-3,5-dinitrobenzotrifluoride (the precursor for the commercial trifluralin herbicides), requires very harsh conditions (110°C, 20+% added SO_3 to enhance the acidity of the medium).

Chlorination or bromination of benzotrifluoride affords rather poor regioselectivity. Even in the presence of $SbCl_5$, chlorination gives a mixture of the three monochloro isomers. Bromination in the presence of Lewis acids is also unselective, with a typical isomer ratio being o:m:p = 26:67:7.

The intrinsic limitations of the range of starting materials that can be used in the industrial chlorination-fluorination route to benzotrifluorides (see below; groups such as NH_2 and OH are not compatible with an HF medium, bromo and iodo groups are liable to displacement by chlorine, and some other substrates such as nitrated toluenes are reluctant to undergo radical chlorination) makes effective and selective methods for subsequent derivatization of benzotrifluoride itself of particular importance and it seems likely that considerable research effort will go into this area.

The presence of CF_3 groups on aromatic rings generally increases their oxidative stability. Thus, in the oxidation of trifluoromethylnaphthalenes, where the CF_3 groups are only on one ring, the only product observed is that resulting from oxidation of the nontrifluoromethylated ring (Figure 5.3).[19]

The trifluoromethyl group is often considered to be relatively inert but it can be vulnerable to attack and is subject to different decomposition pathways.[1] The hydrolytic decomposition of the CF_3 group is substrate dependent, being typically

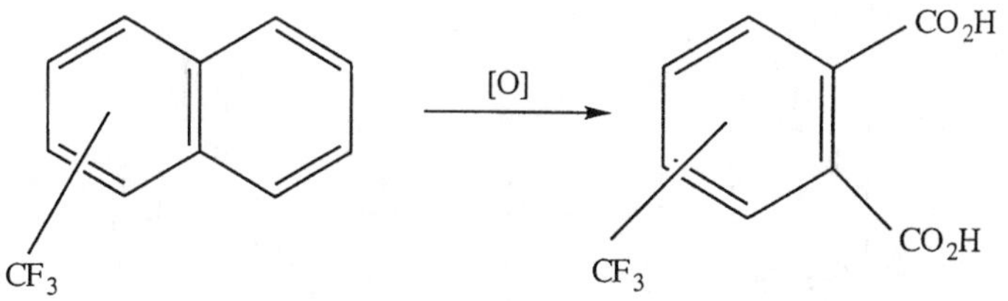

FIGURE 5.3   Oxidation of trifluoromethylated naphthalenes.

base catalyzed in aliphatic systems but normally requiring acid-catalysis in aromatics (Figure 5.4).[20] In contrast, 2-trifluoromethylimidazoles undergo facile hydrolysis in basic media (Figure 5.5).[21,22] The mechanism is believed to involve $-CF_2OH-$ and $-CF=O-$ substituted intermediates. Base-catalyzed hydrolytic decomposition can be achieved if the substrate possesses a sufficiently acidic hydrogen to enable reaction to proceed via elimination of hydrogen fluoride. The ability to lose fluoride ions can be exploited in the synthesis of some compounds, including cyanoimidazoles (via base-catalyzed aminolysis of trifluoromethylated imidazoles)[21,22] and quinolines. The biochemical degradation of trifluoromethyl groups in pharmaceuticals can proceed via a pathway involving loss of fluoride ions and lead to enzyme inhibition.[1,23]

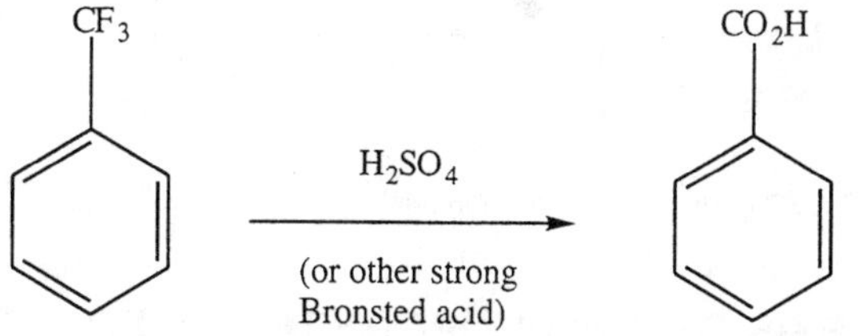

FIGURE 5.4   Acid-catalyzed decomposition of trifluoromethylated aromatics.

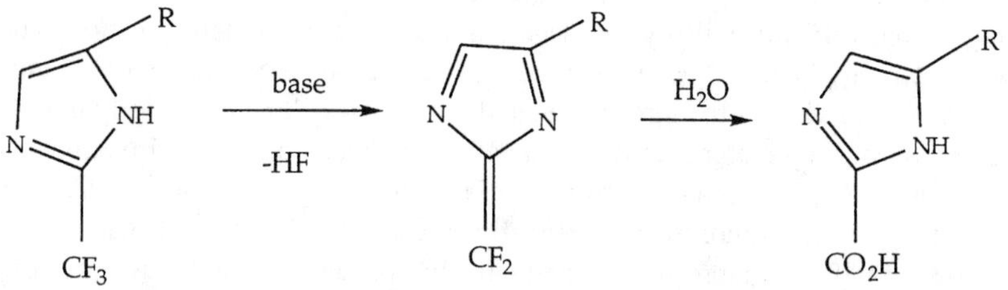

FIGURE 5.5   Base-catalyzed decomposition of trifluoromethylateed imidazoles.

Trifluoromethylated aromatics will react with aluminum chloride via replacement of fluorine by chlorine (Figure 5.6).[24-26] The viability of the halex process effectively precludes the use of this Lewis acid reagent to catalyze Friedel Crafts and other electrophilic reactions on benzotrifluorides. Fluorinated catalysts, notably HF, BF_3, and HF/BF_3, are suitable catalysts for such reactions.

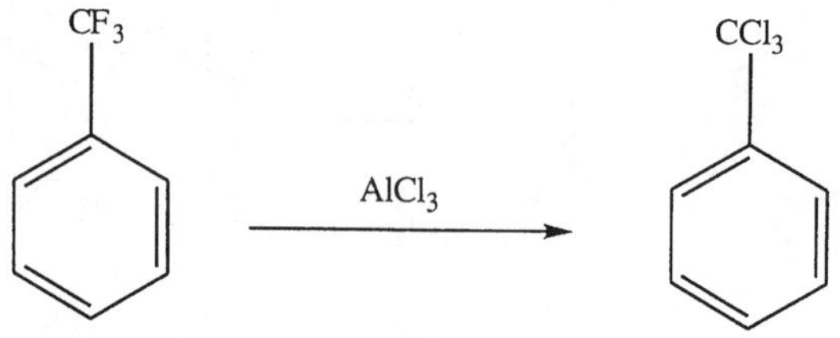

FIGURE 5.6   Reaction of trifluoromethylated aromatics with aluminum chloride.

The trifluoromethyl group can also be reduced to the methyl group by the action of Raney nickel and cobalt alloy in alkaline media (Figure 5.7)[27] and it is conceivable to envisage a route to a desired methylaromatic via the trifluoromethylated analogue, taking advantage of the ability of the CF_3 group to activate substrates to nucleophilic attack (e.g., 2-nitrobenzotrifluoride → 2-fluorobenzotrifluoride → 2-fluorotoluene; this could be considered as a rather expensive alternative to the more familiar Balz-Schiemann route to this product starting from 2-nitrotoluene — see Chapter 3).

FIGURE 5.7   Reduction of trifluoromethylated aromatics to methylaromatics.

5.4   SYNTHESIS OF TRIFLUOROMETHYLATED AROMATICS

The first molecule containing a trifluoromethyl group, benzotrifluoride, was first synthesized in 1898. A very large number of trifluoromethylated aromatics and trifluoromethylated heteroaromatics have been synthesized since then, with the area being driven forward both by the development of new synthetic methods and, as new product applications have emerged, the need to synthesize target molecules.

The preparation of trifluoromethylated aromatics will be discussed on the basis of common starting materials (toluenes, haloaromatics, aromatic hydrocarbons, benzoic acids, and other aromatic substrates). Trifluoromethylated heterocycles can also be built up from trifluoromethylated aliphatics. 1,1,1-Trifluoro-2-penten-4-one,[28] 1-aryl-3,3,3-trifluoro-1-propynes,[29] and hexafluoroacetone[16] can be used as building blocks in this context, although this methodology will not be discussed further.

5.4.1   Routes from Toluenes Using Hydrogen Fluoride and Related Reactions

Swarts's original synthesis of benzotrifluoride used benzotrichloride as the substrate and antimony trifluoride as the fluorinating agent (Figure 5.8).[30,31] It seems likely that the SbF_3 acts as a Lewis acid as well as a source of fluorine. The use of benzotrichloride, normally accessed via the inexpensive reaction of toluene and chlorine, as the substrate is still the basis of many industrial processes, but hydrogen fluoride (again acting as an acid catalyst as well as a source of fluorine) is now preferred to the more expensive SbF_3.[32] A good example of this is the manufacture of 3-aminobenzotrifluoride, a valuable intermediate for dyes, germicides, pharmaceuticals, and crop protection chemicals.[33] The standard manufacturing route is based on a four-step sequence of chlorination, fluorination, nitration, and reduction (Figure 5.9). The process is not selective to the desired product since at the nitrating stage, 2- and 4-nitro isomers are produced (a common problem in aromatic nitration) as side products.[34] Alternative routes to the desired product generally involve the reactions of nitroaromatics at high temperatures: (1) the vapor-phase fluorination of 3-nitrobenzotrichloride (Figure 5.10);[35] and (2) the vapor-phase one-step chlorination-fluorination of 3-nitrotoluene (Figure 5.10).[36]

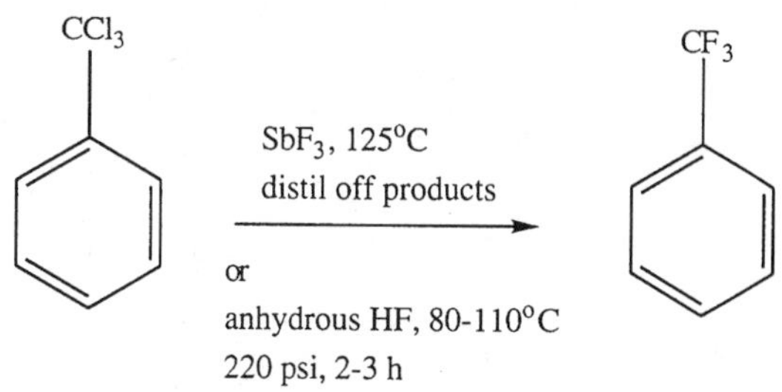

FIGURE 5.8 The conversion of benzotrichloride to benzotrifluoride.

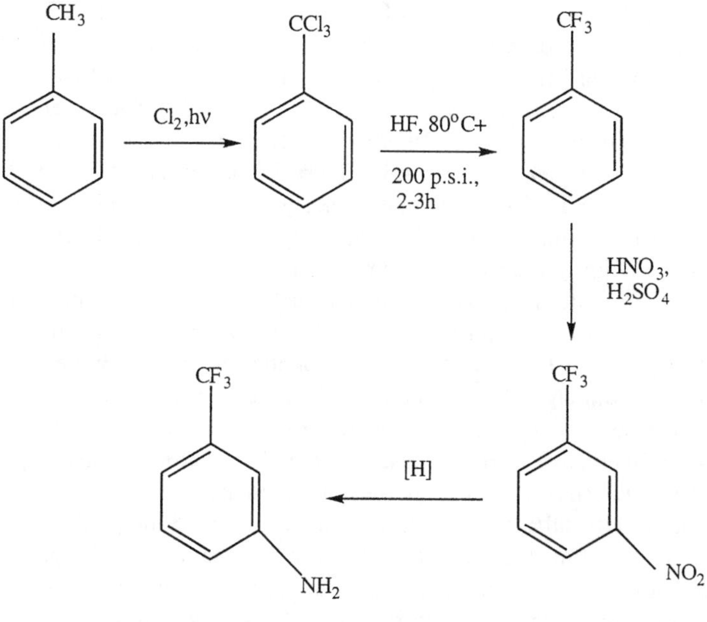

FIGURE 5.9 Standard manufacturing route to 3-aminobenzotrifluoride.

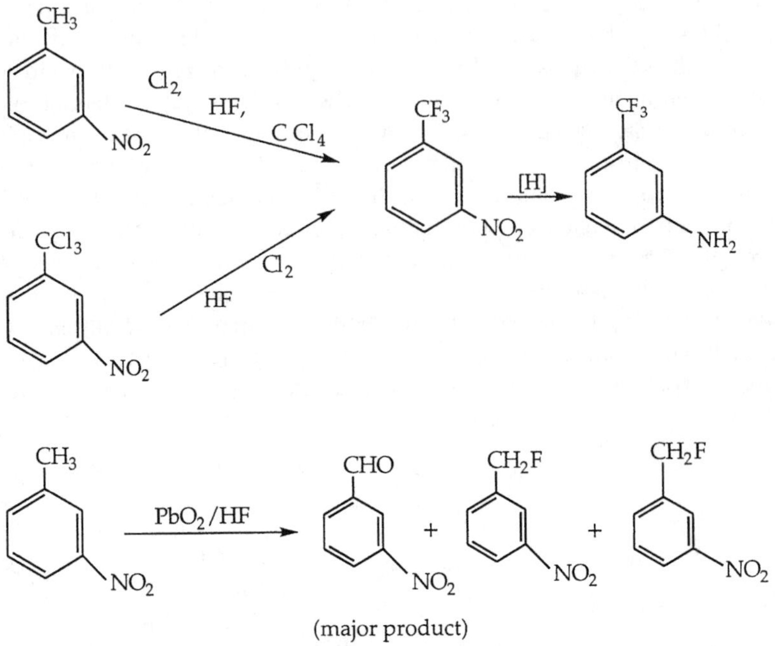

FIGURE 5.10   Alternative routes to 3-aminobenzotrifluoride.

Oxidative fluorination of 3-nitrotoluene gives only partial fluorination at best (Figure 5.10).[37,38] Side-chain fluorination of 3-nitrobenzotrichloride can also be accomplished using ammonium polyfluorides, $NH_4F(HF)_x$,[34] which are easier to handle than HF itself. Remarkably, with high temperatures and pressures over a long period, 3-aminobenzotrifluoride becomes the major product — in high yield (75%) and very good purity on isolation (99.6%). This is probably due to release of iron from the reaction vessel into the reaction system. This will be promoted by the presence of water in the ammonium salt (Figure 5.11). This is not a general method: 2-nitrobenzotrichloride, for example, does not react with ammonium polyfluorides to give the desired 2-aminobenzotrifluoride. Only degradation products are observed on raising the temperature to force the reaction.

The Cl_2/HF route from toluenes to benzotrifluorides is extremely popular and has been applied to the synthesis of N,N-di-n-propyl-4-chloro-2,6-dinitroaniline, "trifluralin" (Figure 5.12), as well as some psychotropic drugs (Figure 5.13). Lewis acids such as antimony (V) compounds can be used alongside HF to aid the fluorination step, especially if moderate temperatures and pressures are preferred.

A likely alternative to routes based on the chlorination of toluenes to benzotrichlorides as intermediates is to use tetrachloromethane as a source of "CCl_3" in the presence of HF and a, typically Lewis acid, catalyst (although it can be possible for the HF to act as an acid catalyst as well as a fluorinating agent, especially when elevated temperatures and pressures can be used to increase the reaction rate). The obvious disadvantage with this approach is the formation of more than one isomer,

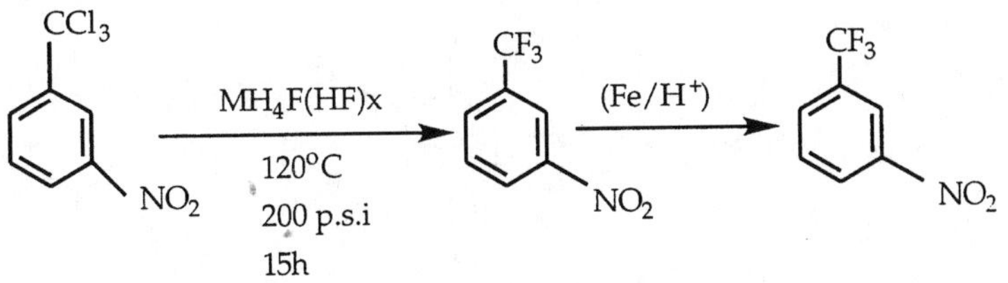

FIGURE 5.11   Reaction of 2-nitrobenzotrichloride with ammonium polyfluorides.

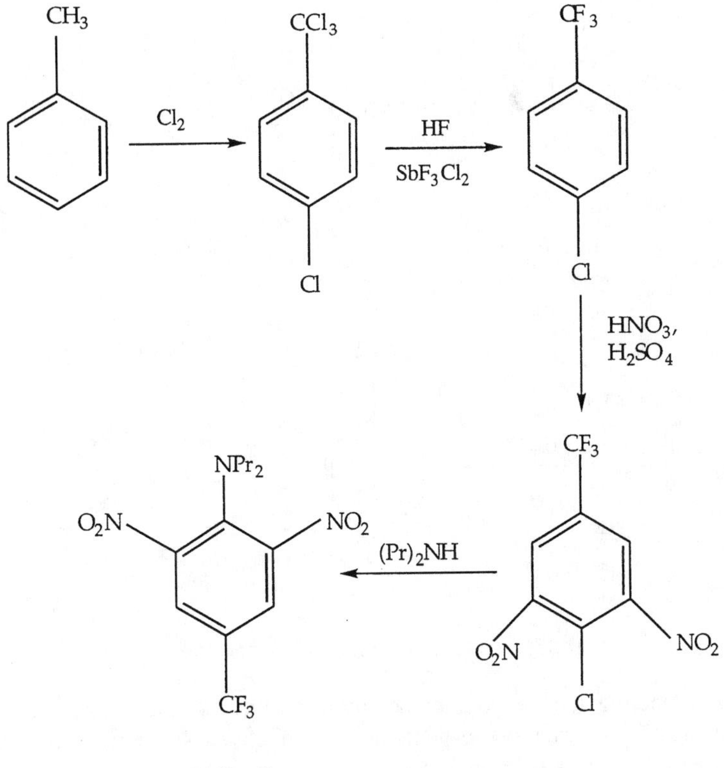

Trifluralin

FIGURE 5.12   Synthesis of Trifluralin.

such as in the synthesis of dichlorobenzotrifluorides from 1,2-dichlorobenzene, for example (Figure 5.14).[1,39]

The preparation of benzotrifluorides from benzotrichlorides using hydrogen fluoride generally requires harsh conditions of temperature and pressure and these, combined with the highly corrosive nature of the acid, create severe operating problems. Alternative reagents are available, such as aluminium trichloride/fluoro-trichloromethane[24-26] and silver tetrafluoroborate.[40] The latter is prohibitively expensive and generally results in only moderate product yields, whereas the former, while

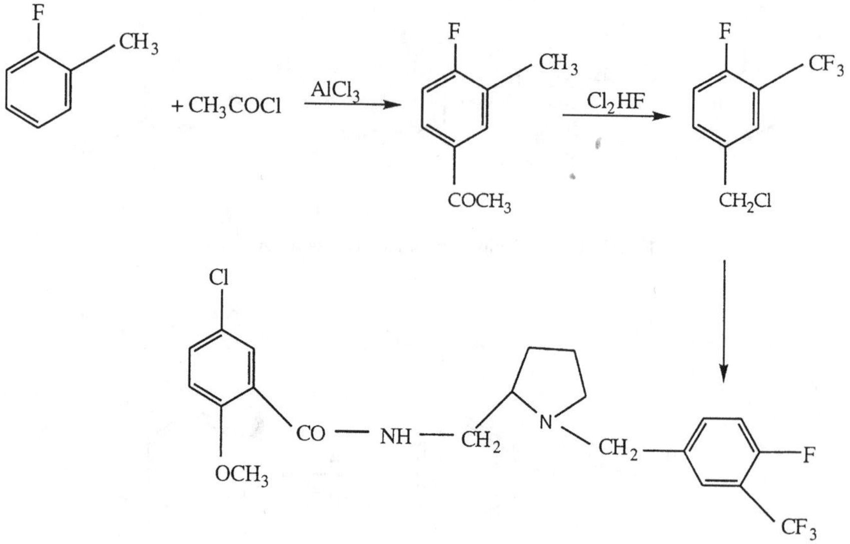

FIGURE 5.13 Synthesis of psychotropic drugs.

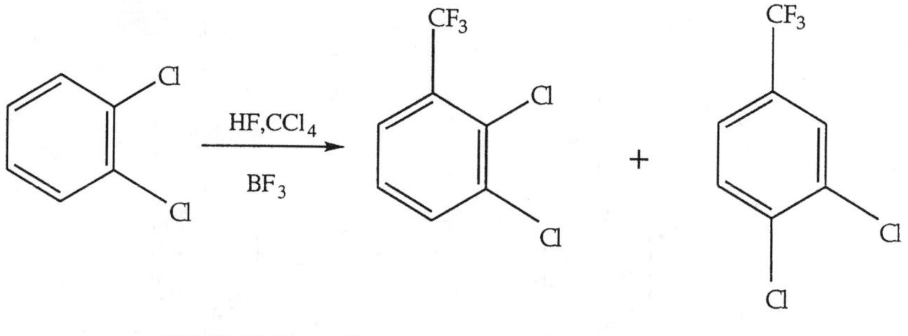

FIGURE 5.14 Trifluoromethylation of 1,2-dichlorobenzene.

capable of reacting at only room temperature, involves two increasingly undesirable (environmentally unfriendly) reagents and its effectiveness seems limited to poly-chlorobenzotrifluorides (Figure 5.15).

5.4.2 Routes from Haloaromatics Using Organometallic Reagents

The coupling reaction of perfluoroalkyl iodides and aryl iodides in the presence of copper was first reported in 1969.[41] Thus, iodobenzene will react with iodotrifluo-romethane in the presence of copper metal to give benzotrifluoride (Figure 5.16). It was thought that the reactive species was "CuCF$_3$" although no compound of that formula has been isolated, and it was 18 years before definitive spectroscopic evidence for its existence in solution was put forward (see Section 5.5).

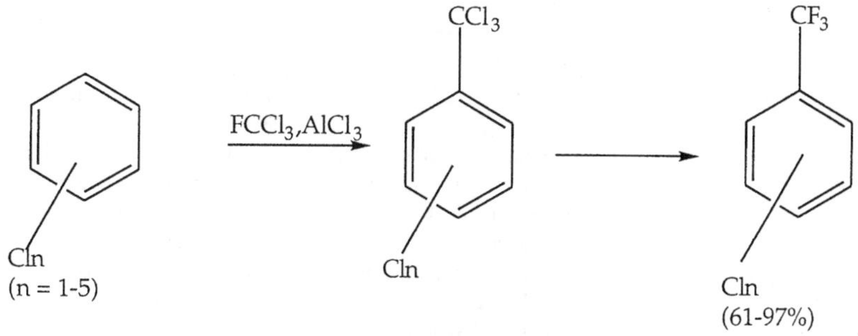

FIGURE 5.15 Reaction of polychlorobenzenes with aluminum chloride/fluorotrichloromethane.

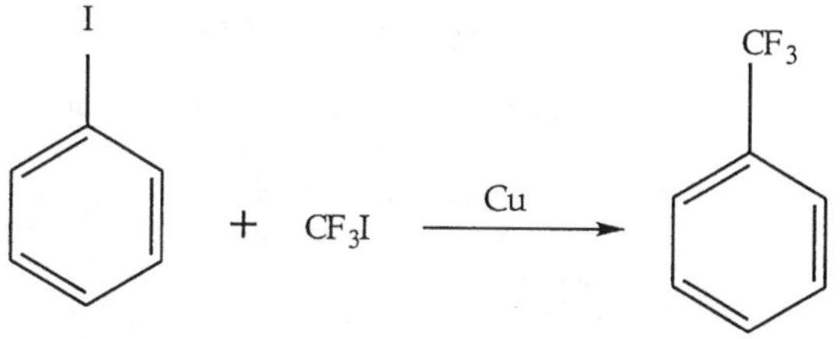

FIGURE 5.16 The reaction of iodobenzene and iodotrifluoromethane in the presence of copper.

There are numerous methods for generating "CuCF$_3$" in solution (Table 5.5), some being rather less than obvious and many involving *in situ* formation of the trifluoromethylating agent. There are definite advantages in preforming the reactive species via the original method of reacting copper metal and iodotrifluoromethane, removing the excess metal, and then adding the aromatic substrate (milder reaction conditions and generally cleaner product mixtures; see Section 5.5). However, cost and convenience factors may persuade the chemist to use *in situ* systems such as copper (I) iodide/trifluoroethanoic acid or the more suprising and mechanistically complex copper metal/dibromodifluoromethane/amide reaction system.[47]

Reactions using "CuCF$_3$" are generally carried out using iodoaromatics as substrates, and these can lead to excellent yields of a range of trifluoromethylated aromatics, although yields can be lower when employing the more thermally demanding methods of generating the copper reagent (Figure 5.17). The methodology can be applied to the synthesis of target molecules such as 2,6-bis(trifluoromethyl)phenol, the key intermediate in the synthesis of some metabolism-resistant analogues of tebufelone, the anti-inflammatory drug (Figure 5.18).

Substitution of bromine and chlorine by the trifluoromethyl group using "CuCF$_3$" is normally much more difficult. In the simplest case, bromobenzene will react with many "CuCF$_3$" reagents, but the rate of reaction is slow and the final yield can be poor. This is a result of decomposition of the reagent, copper-catalyzed side reactions, and, in cases where the CF$_2$ carbene may be present, chain extension leading to perfluoroalkylation.

Table 5.5 Methods of Generating "CuCF$_3$"

Method	Comments
Copper metal + CF$_3$I[42]	Original method still preferred when expense is not a factor; can be preformed at low temperatures (avoiding chain extension) and used with no excess copper, hence avoiding many side reactions
Copper (I) iodide + CF$_3$CO$_2$H[43]	Requires high temperatures for decarboxylation to occur (>100°C); thought to involve [CuCF$_3$I]$^-$; relatively inexpensive, but often rather inefficient
Copper metal + Hg(CF$_3$)$_2$[44]	Enables preformation of "CuCF$_3$" but mercury toxicity is a disadvantage
Copper metal + CF$_3$N(NO)SO$_2$CF$_3$[45]	Requires presynthesis of reagent from trifluoronitrosomethane, hydroxylamine, and trifluoromethanesulfonyl chloride
Electrochemical using copper anode, gaseous CF$_3$Br plus other agents and solvent	Relatively inexpensive reagent but complex, energy-intensive chemistry
Copper (I) salts + Cd(CF$_3$)$_x$X$_y$ and Zn(CF$_3$)$_x$X$_y$ (x = 1 or 2)[48]	Quite straightforward once organometallic reagent is available
Copper metal + CF$_2$Br$_2$ + amide (typically as the solvent)[49]	Reactions believed to involve initial formation of CF$_2$ carbene; interesting alternative to normal preformed CF$_3$ precursor; requires amide reactant, which is also commonly used as the solvent
Copper (I) iodide + FO$_2$SCF$_2$CO$_2$Me[50]	Reagent is easy to use and reasonably available from its use to prepare Nafion–H
Copper (I) iodide + FSO$_2$CF$_2$CF$_2$OCF$_2$CO$_2$Me[51]	Provides a remarkably reactive form of "CuCF$_3$" capable of attacking chloroaromatics
Copper (I) iodide + KF + ClCF$_2$CO$_2$Me[52]	*In situ* halex accompanying decarboxylation; high temperatures (>100°C) required
Copper (I) iodide + CF$_3$SiMe$_3$[53]	Reactions with iodoaromatics generally require the presence of F$^-$

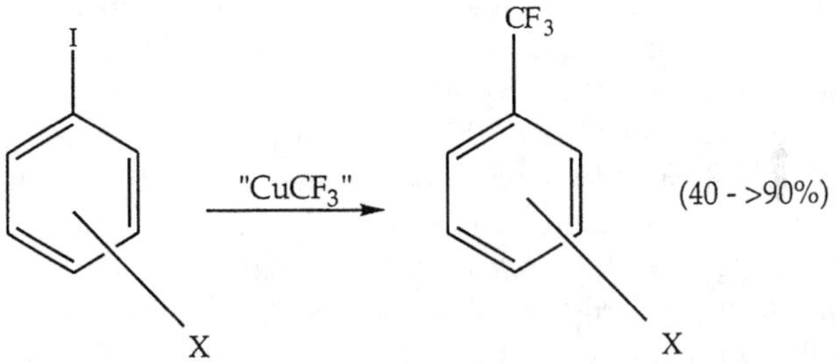

X = H, o-NO$_2$, m-NO$_2$, p-NO$_2$, o-CH$_3$, m-CH$_3$, p-CH$_3$, p-Cl, p-Br, p-OMe

FIGURE 5.17 The reaction of "CuCF$_3$" with iodoaromatics.

Better product yields from bromo- and even chloroaromatic substrates can be achieved in some cases, notably in reactions involving:

1. Heteroaromatic substrates
2. Chelating substrates
3. Highly reactive substrates

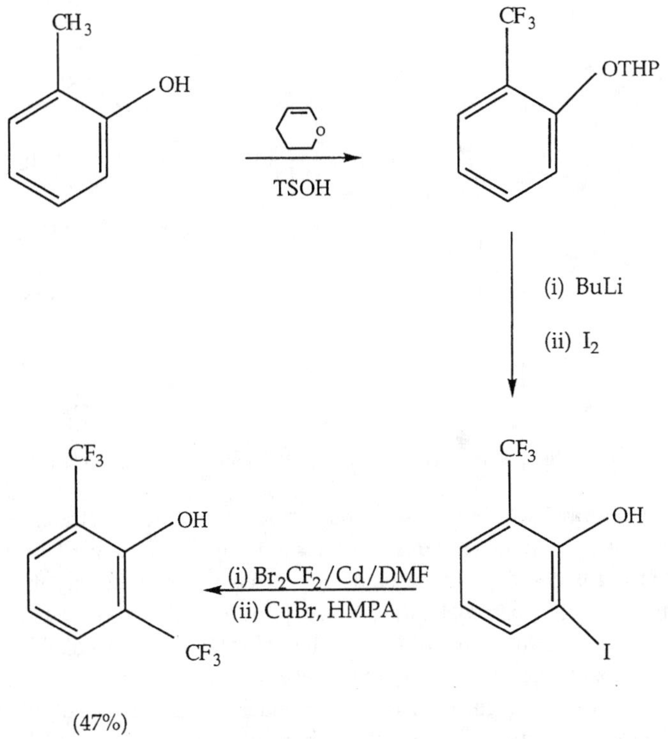

FIGURE 5.18 Synthesis of 2,6-bis(trifluoromethyl)phenol from 2-hydroxybenzotrifluoride acid.

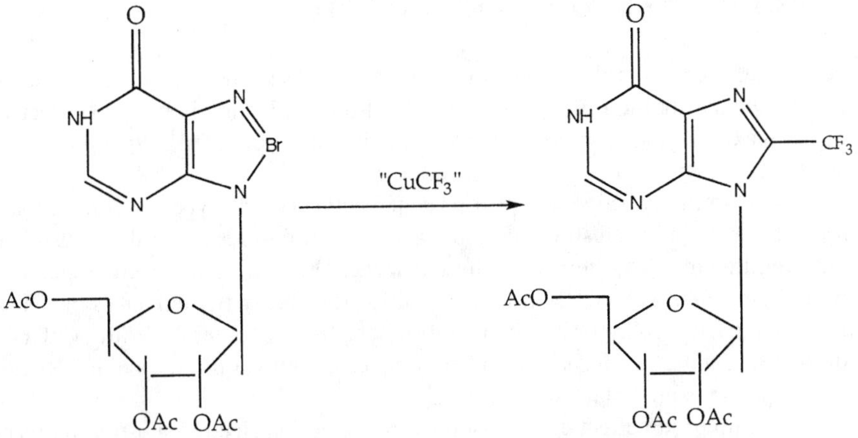

FIGURE 5.19 Brominated nucleotide in trifluoromethylation using "CuCF$_3$".

Brominated nucleotides can be used as starting materials for the synthesis of the trifluoromethylated analogues (Figure 5.19). In some cases this may even be extended to chlorinated nucleotides (Figure 5.20).

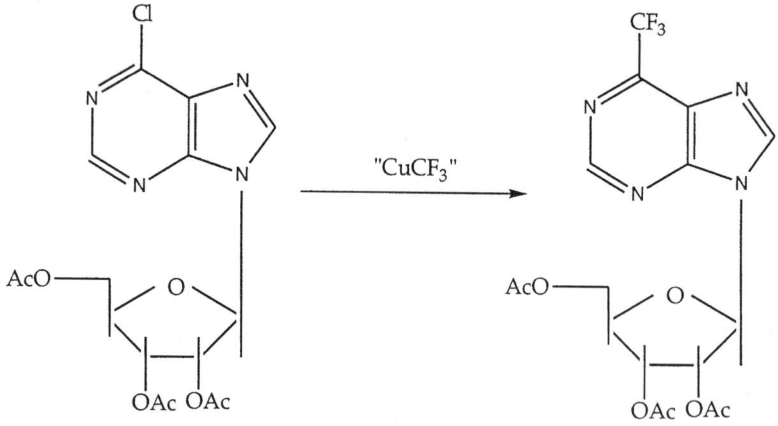

FIGURE 5.20   Chlorinated nucleotide in trifluoromethylation using "CuCF$_3$".

The major drawbacks with using iodoaromatics as the substrates in trifluoro-methylations using copper reagents are the high cost, limited availability of more complex starting materials, and the facile copper-catalyzed coupling of iodoaromatics to biphenyls. Chloroaromatics are much preferred but are much less reactive, unless they possess a substituent ortho to the chlorine which is capable of holding the copper reagent in close proximity to the reaction site.[54,56] Thus, nitro and carbonyl groups are effective, ring nitrogens less so (hence the possible use of chlorinated and brominated nucleotides as described above), whereas cyano and alkyl groups are ineffective (Figure 5.21). This chelating effect can be illustrated by the relative reactivity of nitroaromatic substrates towards "CuCF$_3$":

$$2\text{-NO}_2\text{-C}_6\text{H}_4\text{Cl} > 4\text{-NO}_2\text{-C}_6\text{H}_4\text{Cl} > 3\text{-NO}_2\text{-C}_6\text{H}_4\text{Cl}$$

which is different from the order of reactivities of these substrates in conventional nucleophilic substitution reactions where the 4-nitro substrate is the most reactive due to maximized mesomeric effects (e.g., in nucleophilic fluorination — see Chapter 2).

The dissimilarity between copper reagent reactions of this type and conventional aromatic nucleophilic substitutions using alkali metal fluorides is further illustrated by the reaction of 2,3,5,6-tetrachloronitrobenzene. This is a highly reactive substrate towards nucleophilic fluorodenitration, the nitro group being twisted out of the plane of the ring by the bulky ortho chloro-substituents (see Chapter 2). With "CuCF$_3$", however, no denitration occurs and only a complex mixture of trifluoromethyldechlo-rination products are obtained (Figure 5.22).[56]

The ortho group effect can be usefully exploited in selective substitution reactions. Thus, 2,3-dichloronitrobenzene reacts with "CuCF$_3$" to give 6-chloro-2-nitrobenzotrifluoride in good yield, with no contamination from the other isomer or the bis(trifluoromethyl) product (Figure 5.23).[54]

When "CuCF$_3$" is generated *in situ* from the copper/dibromodifluoromethane/amide reagent system, contamination from perfluoroalkyl products can be a problem,

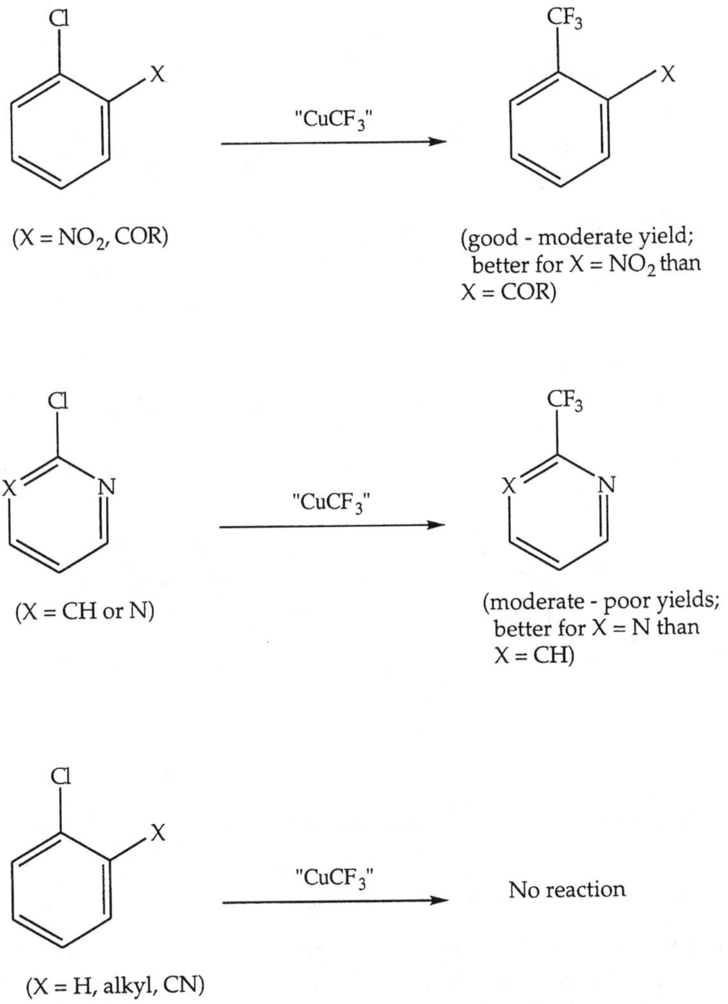

FIGURE 5.21 Ortho-substituent-dependent trifluoromethylation of chloroaromatics using "CuCF$_3$".

especially with the less active substrates, and they can even become the major products.[56] The yield of the desired product can be enhanced in some cases by the *in situ* use of highly dry, activated charcoal.[55]

The copper (I) iodide/methyl-3-oxo-w-fluorosulfonylperfluoropentanoate reagent system acts as an *in situ* source of "CuCF$_3$" and shows remarkably high reactivity towards a broad range of substrates, including iodoaromatics, bromoaromatics, and even chloroaromatics. Thus, chloroaromatics containing one electron-withdrawing group (NO$_2$, CHO, CO$_2$Et) give good yields of the desired trifluoromethylated products, and the presence of electron-donating groups (Me, MeO) only has a small effect on the product yields, which remain reasonable (Table 5.6). The remarkably high reactivity of this less than obvious reagent system may be due to a high dynamic concentration of the key precursors, F$^-$ and the CF$_2$ carbene.[51]

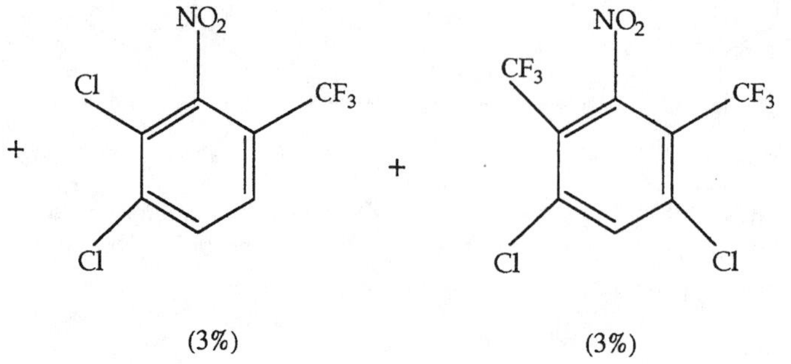

FIGURE 5.22   Reaction of 2,3,5,6-tetrachloronitrobenzene with "$CuCF_3$".

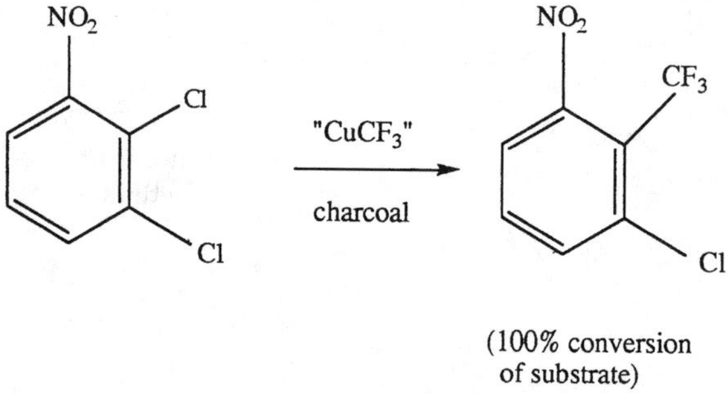

FIGURE 5.23   Selective monotrifluoromethylation of 2,3-dichloronitrobenzene using "$CuCF_3$".

5.4.3 Routes from Aromatic Hydrocarbons

The direct trifluoromethylation of aromatic hydrocarbons involves the substitution of H by CF_3 and can be accomplished either with the use of trifluoromethyl radicals or trifluoromethyl cations. Trifluoromethyl radicals have been known since 1948 and they can be generated by photochemical, electrochemical, thermal, or chemical reactions.[1] A summary of the main trifluoromethyl radical precursors is given in Table 5.7.

Trifluoromethyl radicals are electrophilic and should readily react with electron-rich aromatics, including anilines, acetanilides, anisoles, phenol, and even benzene itself (Table 5.8). Several important points should be noted:

1.   Yields can vary significantly and are dependent on the reagent and the substrate. For benzene, the best results are obtained using iodotrifluoromethane and mercury (Table 5.9).[57]
2.   Trifluoromethyl radicals are not very specific, although they do show preference for sites of high electron density in the highest occupied molecular orbital (HOMO).

Table 5.6 Reactions of Chloroaromatics with Methyl 3-oxo-w-fluorosulfonyl-perfluoropentanoate and Copper (I) iodide in DMF

Substrate	Conditions t/h; T°C	Yield of Trifluoromethylated Product (%)
Cl — (phenyl)	8; 120	52
Cl — (phenyl) — NO$_2$	8; 120	67
Cl — (phenyl) — CHO	8; 120	56
Cl — (phenyl) — CH$_3$	8; 120	32
Cl — (phenyl) — OMe	8; 120	37

3. Several heteroaromatics react to give only one product, including pyrroles (Figure 5.24),[58-60] uracil (Figure 5.25),[73-76] thiophene,[71] and 2-substituted imidazoles.[77-79] Other heteroaromatics give a mixture where the isomer distribution can be highly reagent dependent. These include pyridines (Figure 5.26),[58,66,80,81] imidazole (Figure 5.27),[77-79] and benzothiophene.[71] In complex molecular systems, the imidazole ring been shown to be especially prone to photochemical trifluoromethylation.[82]

S-, Se-, and Te-trifluoromethylated dibenzoheterocyclic onium salts, as well as the trifluoromethyldibenzothiophenium salts, can be used as sources of the trifluoromethyl cation in the direct trifluoromethylation of activated aromatics and heteroaromatics.[83,84] Their reactivity depends on the heteroatom (S > Se > Te) and on the

Table 5.7 Trifluoromethyl Radical Precursors

Reagent	Experimental Conditions
CF_3CO_2Ag	TiO_2 as photocatalyst/irradiate
$(CF_3CO_2)_2$	Electron donation
CF_3CO_2H	Electrolysis or XeF_2
CF_3COCF_3	Heat or irradiate
CF_3SO_2Na	t-BuOOH
$CF_3N(NO)SO_2CF_3$	Heat or irradiate
$CF_3N(NO)SO_2Ph$	Irradiate
$CF_3N=NCF_3$	Irradiate
$Hg(CF_3)_2$	Irradiate
CF_3Br	Heat or irradiate
CF_3I	Heat or irradiate

Table 5.8 Reactions of Aromatics with Sources of Trifluoromethyl Radicals

Substrate	Reaction Conditions	Yield (%) and Selectivity (o:m:p)	Ref.
Cl (phenyl ring)	CF_3I, 198°C	50%; 2.1: 1: 1.2	61
Br (phenyl ring)	CF_3I, Heat	61%; 48: 30: 22	62
Me (phenyl ring)	CF_3CO_2Ag/TiO_2/irradiate,	38%; 2: 1: 1	63
OH (phenyl ring)	CF_3SO_2Na/t-BuOOH/Cu (II)	45%; 4 : 1: 6	64
NH_2 (phenyl ring)	$CF_3Br/Zn/SO_2$	56%; 1.8: 0: 1	65

Table 5.9 Relative Reactivities of Different Sources of Trifluoromethyl Radicals Towards Benzene

CF$_3$ Source	Experimental Conditions	Yield of PhCF$_3$ (%)	Ref.
CF$_3$Br	Na$_2$S$_2$O$_4$	17	64, 66
CF$_3$I	Irradiate/Hg	65	57
CF$_3$I	Irradiate	14	58
CF$_3$CO$_2$H	XeF$_2$	33	67, 68
(CF$_3$)$_2$Hg	150°C	51	69, 70
(CF$_3$)$_2$Te	Irradiate	20	69, 70
(CF$_3$)$_2$Te	150°C	31	69, 70
(CF$_3$CO$_2$)$_2$	70°C	54	71
CF$_3$CO$_2$Hg	TiO$_2$/irradiate	50	72

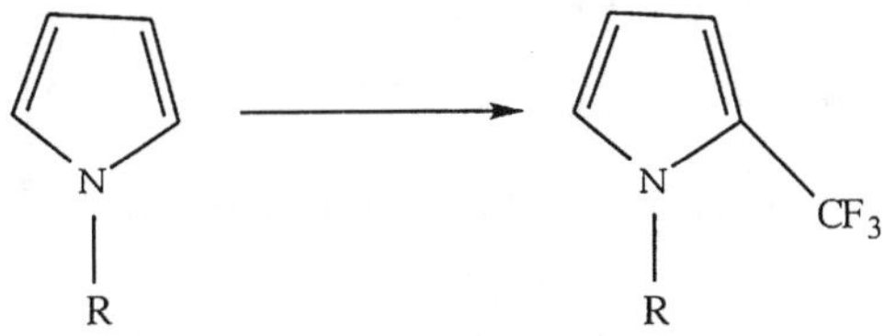

R	Experimental Conditions	Yield (%)
Me	CF$_3$Br/hν	6.5
	CF$_3$I/hν	35
	CF$_3$Br/Zn/SO$_2$	52
PhCH$_2$	CF$_3$I/hν	60
H	CF$_3$I/hν	33
	(CF$_3$CO$_2$)$_2$	56

FIGURE 5.24 Reactions of pyrroles with bifluoromethyl radical sources.

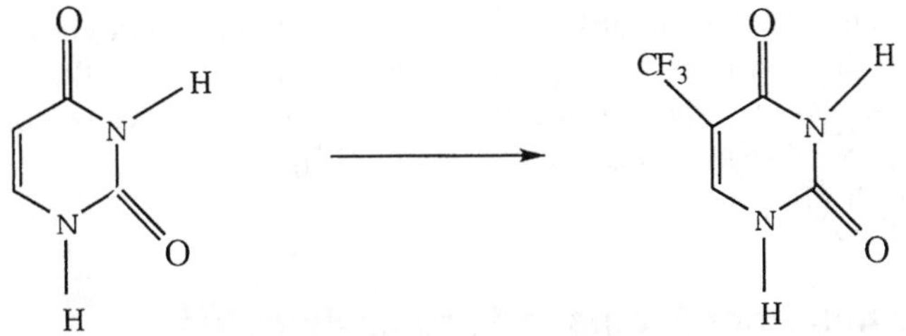

Experimental Conditions	Yield (%)
Hg(CF$_3$)$_2$	> 60
CF$_3$CO$_2$H/electrolysis	60
CF$_3$Br/hν	11

FIGURE 5.25 Reactions of uracil with sources of bifluoromethyl radicals.

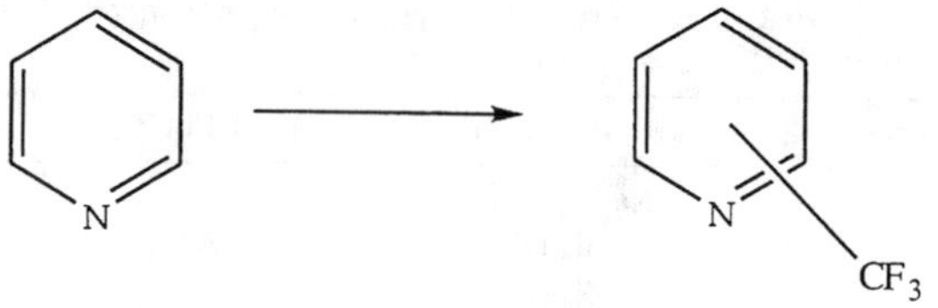

Experimental Conditions	Yield (%)	Selectivity (o : m : p)
CF₃Br/hv	27	-
CF₃Br/Zn/SO₂	8	5 : 3 : 1
CF₃CO₂H/electroysis	7	27 : 38 : 35
CF₃I/hv	81	38 : 27 : 16

FIGURE 5.26   Reactions of pyridine with sources of trifluoromethyl radicals.

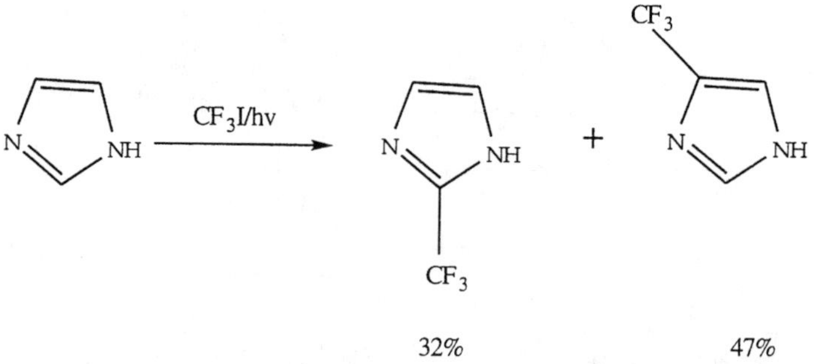

FIGURE 5.27   Reactions of imidazole with a source of trifluoromethyl radicals.

substitution on the rings (dinitro substitution being especially effective) so that the reagents can be considered as "power-variable electrophilic trifluoromethylating agents". Good yields of the monotrifluoromethylated products can be obtained with activated benzenoid substrates such as aniline (Figure 5.28) and heteroaromatics such as pyrrole (Figure 5.29).

5.4.4 Routes from Aromatic Carboxylic Acids

The conversion of a carboxylic acid function to a trifluoromethyl group using sulfur tetrafluoride or related reagents is a well-known synthetic method in organofluorine chemistry which can be readily applied to aromatic systems (Figure 5.30).[1,85,86] Electron-withdrawing groups on the aromatic ring cause a reduction in product yield, although this problem can, at least partly, be remedied by the concomitant use of hydrogen fluoride. Methyl esters of aromatic carboxylic acids are also viable substrates. The method can be readily extended to heterocyclic substrates such as thiazoles (Figure 5.30). Other groups such as alkyl and halogen are not affected.

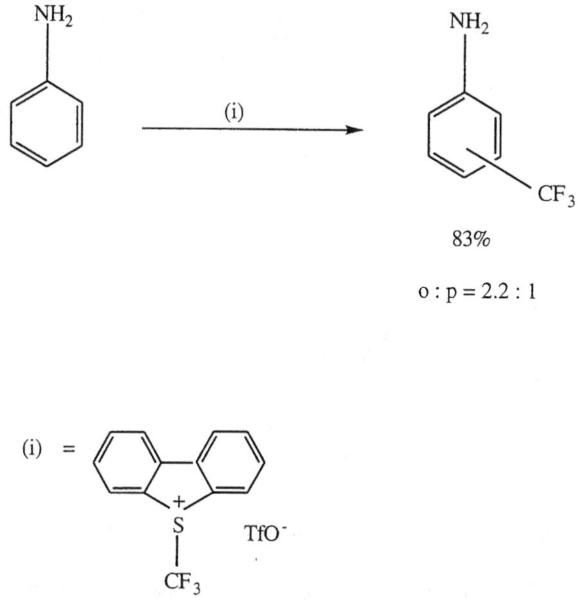

FIGURE 5.28 Trifluoromethylation of aniline using trifluoromethyldibenzothiophenium triflate.

The intermediate acid fluorides may be observed as impurities in inefficient reaction systems. Reactions normally occur at moderated temperatures and aromatic carboxylic acids are generally available and relatively inexpensive. The major drawbacks with the method are the extreme toxicity and handling difficulties associated with sulfur tetrafluoride along with the prohibitive cost of the reagent. Diethylaminosulfur trifluoride (DAST) is a more user-friendly alternative to sulfur tetrafluoride, it being a liquid (sulfur tetrafluoride is a gas) and less aggressive in its behavior. In combination with sodium fluoride, DAST can be used to convert benzoic acid to benzotrifluoride, for example (Figure 5.30). Unfortunately, DAST is also extremely expensive.

5.4.5 Routes from Other Aromatic Substrates

Some thio groups can be replaced by the trifluoromethyl group. Thus, the $-C(SMe)_3$ group is converted to $-CF_3$ on simple aromatic nuclei by reaction with 1,3-dibromo-5,5-dimethylhydantoin and then hydrogen fluoride in pyridine at low temperatures (Figure 5.31).[87-89] Aromatic dithiocarboxylic acids, which are conveniently prepared by reaction of the aromatic Grignard reagent with carbon disulfide,[90] react with xenon difluoride to give moderate yields of benzotrifluorides (Figure 5.31). Alternatively, trifluoromethyl-substituted aromatic compounds can be obtained by the oxidative desulfurization-fluorination of arenedithiocarboxylate esters (which can be prepared from the carboxylic acids) using a combination of 1,3-dibromo-5,5-dimethylhydantoin and a polyfluoride (Figure 5.31),[91] or by using bromine trifluoride in trichlorofluoromethane (although this may lead to the formation of some $ArCF_2Cl$ due to a side reaction of the BrF_3 and the solvent).[92] Interestingly, replacement of the hydantoin by N-bromosuccinimide results in the formation of partially fluorinated difluoro(methylthio)-substituted aromatics.[91]

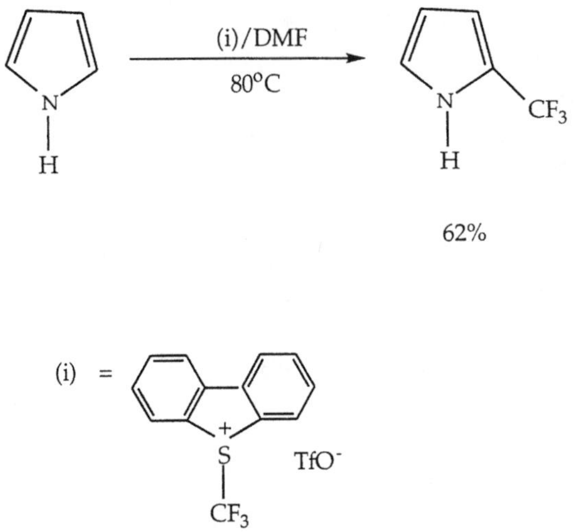

FIGURE 5.29 Trifluoromethylation of pyrrole using trifluoromethyldibenzothiophenium triflate.

Aromatic trifluoroacetates can be decomposed to the trifluoromethylaromatics by flash thermolysis, although the yields are low and mixtures of products can occur (Figure 5.32).[1]

5.5 SYNTHETIC METHODS

The most commonly used fluorine reagents used in the synthesis of trifluoromethyl-substituted aromatics are given in Table 5.10, along with guidance on availability, handling, and use. The original preparation of benzotrifluoride from benzotrichloride was carried out using SbF_3, the least reactive of the antimony-based fluorinating agents. This can be contrasted with the analogous preparation of trifluoromethyl-substituted aliphatics, which require the more reactive Sb (V) compounds such as SbF_3Br_2. Anhydrous hydrogen fluoride is actually capable of converting benzo-trichloride to benzotrifluoride when the reaction is carried out in an autoclave (capable of withstanding attack by HF) at 40°C. The combination of Sb (III) and HF is more reactive in such systems.

Conventional nucleophilic fluorinating agents that are effective in converting benzyl chloride to benzyl fluoride (e.g., the more reactive alkali metal and onium fluorides — see Chapter 2) are not suitable for the preparation of the trifluoromethylated analogues, presumably because of the sterically crowded methyl carbon being inaccessible to the fluoride ion or ion pair. Rather, reactions require acid catalysis and the mechanism presumably involves removal of Cl⁻ by H⁺ or a Lewis acid center to produce the highly reactive $ArCCl_2^+$ cation which will be immediately quenched by F⁻ (Figure 5.33).

Trifluoromethylcopper (I) has not been isolated and evidence for its existence is largely based on low-temperature solution nuclear magnetic spectroscopy. It can, however, be generated in several different ways starting from copper metal or a copper (I) salt (typically the iodide). There are several advantages in preforming the

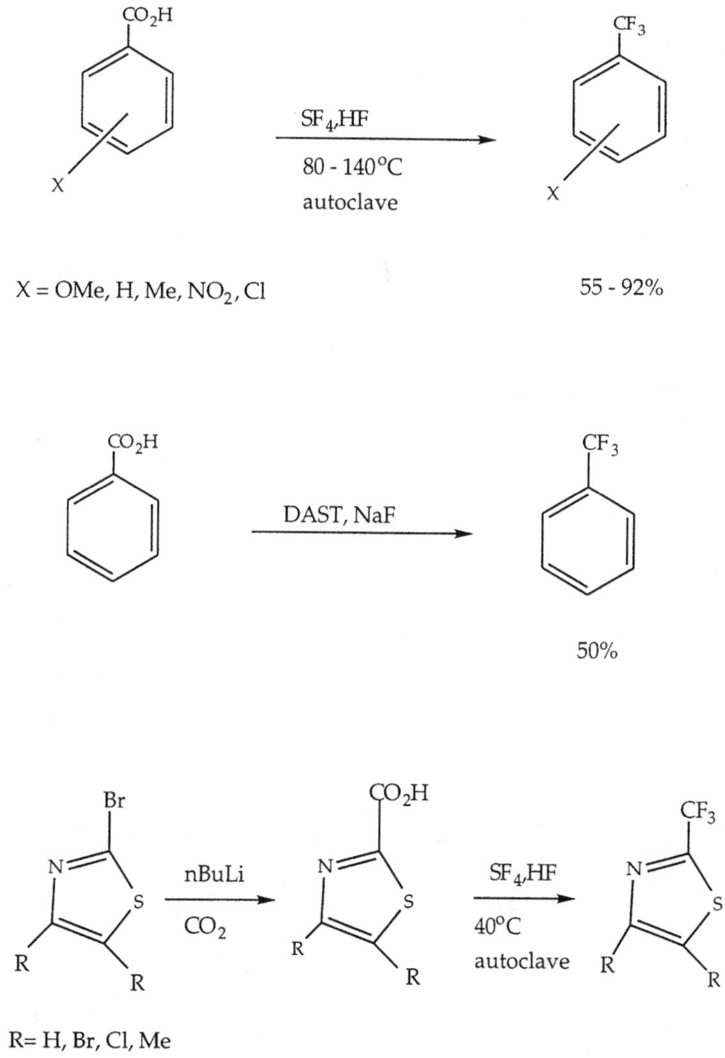

FIGURE 5.30 Conversion of aromatic carboxylic acids to trifluoromethyl-substituted aromatics using sulfur tetrafluoride.

"CuCF$_3$," from iodotrifluoromethane and copper metal and removing the excess metal.[42] These include the use of mild reaction conditions in subsequent trifluoromethylations and the reduction in side products such as biaryls, which can be easily formed in reactions involving copper metal. Solutions of "CuCF$_3$," should be handled in an inert atmosphere since the reagent decomposes on exposure to air. Slow conversion of "CuCF$_3$," to "CuC$_2$F$_5$," can also occur and subsequently lead to the formation of chain extension products. This is a serious problem when the reagent is generated using a CF$_2$-precursor such as CF$_2$Br$_2$,[49] since the mechanism for this reaction involves CF$_2$ carbenes, which can easily add to "CuCF$_3$," if the reagent does not immediately react with the aromatic substrate.

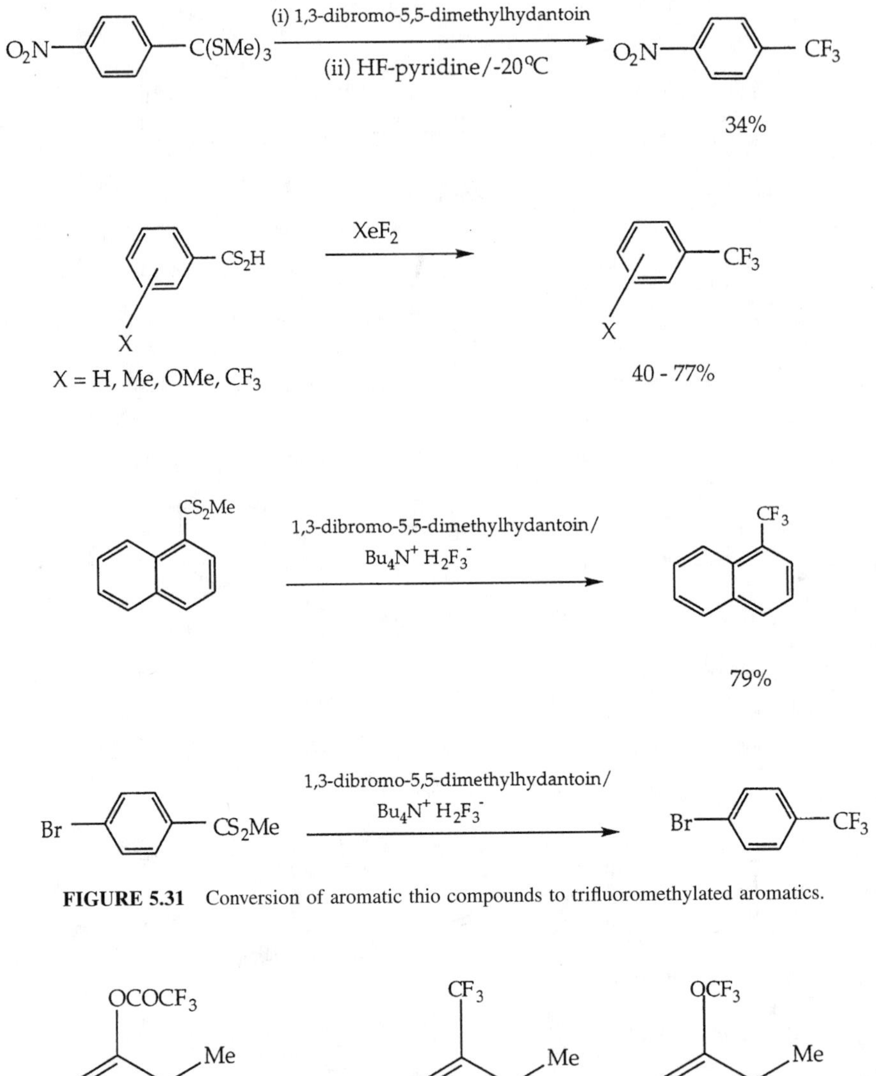

$$O_2N-\langle \rangle-C(SMe)_3 \xrightarrow[\text{(ii) HF-pyridine/-20°C}]{\text{(i) 1,3-dibromo-5,5-dimethylhydantoin}} O_2N-\langle \rangle-CF_3$$

34%

$$\langle \rangle-CS_2H \xrightarrow{XeF_2} \langle \rangle-CF_3$$

X   X

X = H, Me, OMe, CF_3   40 - 77%

CS_2Me   CF_3

1,3-dibromo-5,5-dimethylhydantoin/
Bu_4N^+ H_2F_3^-

79%

$$Br-\langle \rangle-CS_2Me \xrightarrow[\text{Bu}_4N^+ H_2F_3^-]{\text{1,3-dibromo-5,5-dimethylhydantoin/}} Br-\langle \rangle-CF_3$$

FIGURE 5.31 Conversion of aromatic thio compounds to trifluoromethylated aromatics.

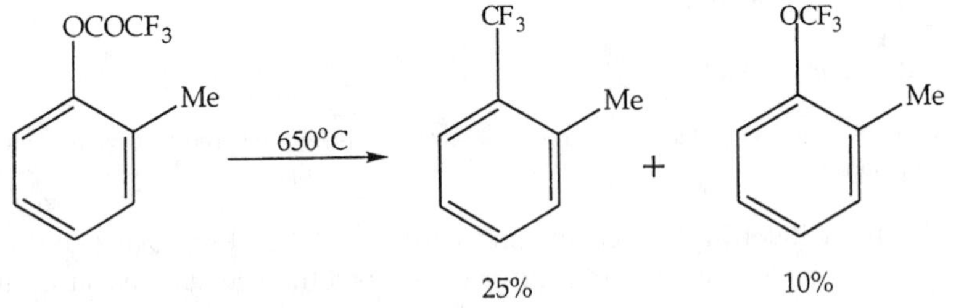

OCOCF_3   CF_3   OCF_3

Me   Me   Me

650°C   +

25%   10%

FIGURE 5.32 Decomposition of aromatic trifluoroacetates.

Trifluoromethylations using "CuCF_3" are thought to involve copper-assisted nucleophilic substitution. Of the common aromatic substituents, only OH and NH_2 groups seriously inhibit reaction of iodoaromatics but bromo- and chloroaromatics usually require the presence of an ortho group capable of coordinating to copper

Table 5.10 Major Fluorine Reagents Used in the Synthesis of Trifluoromethylated Aromatics

Reagent	Availability/Preparation	Comments on Handling and Use
HF	Commercially available as anhydrous acid, aqueous solutions, HF-pyridine complex, and as polyfluorides	Anhydrous HF is a strong Brönsted acid and may cause acid-catalyzed side reactions; anhydrous HF and concentrated solutions are corrosive and must be handled with extreme care; HF attacks glass
SbF_3 and other antimony halides	Commercially available as a crystalline solid or prepared by action of HF on Sb_2O_3	Solid sublimes at >300°C; highly soluble in water; Sb(V) fluoride and mixed halides are extremely powerful Lewis acids and should be handled with care
"$CuCF_3$"	Not available commercially, cannot be isolated; prepare immediately before reaction or *in situ* using methods given in Table 5.5	Excess copper metal can cause unwanted side reactions; solutions are air sensitive and must be handled in an inert atmosphere; slow conversion to CuC_2F_5 can also occur
Trifluoroethanoic acid and its salts	Commercially available (acid) or easily made from the acid	The acid is corrosive and may cause unwanted acid-catalyzed side reactions
CF_3X (X = I, etc.)	All of the halides are commercially available, as are several other CF_3 reagents	The halides are gases
SF_4	Commerically available (gas) but very expensive; diethylaminosulfur trifluoride (DAST) is a liquid (b.p. 30°C/3 mm) but also very expensive	Highly toxic and corrosive; DAST is also highly corrosive; SF_4 shows amphoteric Lewis acid-base properties and rapidly decomposes in the presence of moisture ($\rightarrow SO_2 + HF$)

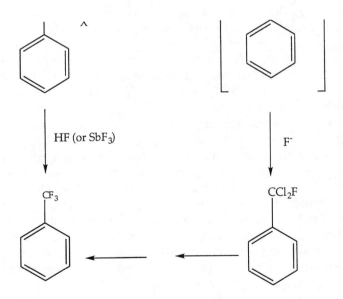

FIGURE 5.33 Conversion of benzotrichlorides to benzotrifluorides.

and hence helping to hold the "CuCF$_3$" in position. Reactions generally employ dipolar aprotic solvents or pyridines and specifically dialkylamides in Cu/CF$_2$Br$_2$ reagent systems. Reaction rates may be increased by the presence of high surface area, well-dried wood charcoal, which may act by stabilizing the "CuCF$_3$".[55] Perfluoroalkylation resulting from chain extension reactions (giving –C$_2$F$_5$ and higher substitutions) may also increase.

Iodotrifluoromethane and bromotrifluoromethane are probably the best known sources of CF$_3$ in trifluoromethylations. The low boiling point of these reagents is a drawback, however, especially when so many aromatic trifluoromethylations require elevated temperatures. Trifluoroethanoic acid or the anhydride are higher boiling analogues and are relatively inexpensive, although they normally only release CF$_3$ on heating in the presence of base (to generate the trifluoroethanoate anion — the salts are commonly used as reagents) or under electrochemical conditions. Hexafluoropropanone is also readily available and relatively inexpensive and can act as a source of CF$_3$ (typically as the radical) under heat or photolysis. Several other sources of CF$_3$ radicals are known (Table 5.7), but generally involve less convenient or more expensive reagents.

Sulfur tetrafluoride, while being an excellent reagent for the direct conversion of the readily available and inexpensive aromatic carboxylic acids to trifluoromethylated analogues, suffers from the major drawbacks of high cost, difficulty in handling (it is a gas), high toxicity, and ready hydrolysis (producing the highly corrosive HF). The most commonly used alternative, diethylaminosulfur trifluoride (DAST) is a liquid and is less easily hydrolyzed. Unfortunately, DAST is also extremely expensive and may not be reactive enough for less reactive aromatic carboxylic acids.

Electrophilic radical routes to benzotrifluorides are mostly limited by the need for highly toxic reagents, high pressures, and various other drawbacks that are likely to preclude the use of this methodology on a large scale. However, the development of more benign methods, such as those based on trifluoroethanoic anhydride and a solid peroxygen source (e.g., sodium percarbonate; this can be used for the *in situ* formation of bis(trifluoroacetyl) peroxide, a useful source of CF$_3$ radicals) may enable better use of this route to benzotrifluorides.[93]

The choice of the best substrate and reagent(s) to obtain a particular trifluoromethylated aromatic product is not trivial. The conversion of ArCCl$_3$ (readily accessed via chlorination of the toluene or *in situ* trichloromethylation of the aromatic using CCl$_4$) to ArCF$_3$ remains popular, especially in industry, where the low cost of the reagents outweighs the disadvantages of using highly corrosive and hazardous HF-based reagents. In the laboratory, organometallic routes and those employing "CuCF$_3$" in particular, are often favored, especially if the iodoaromatic substrate is readily available. Yields are, however, often less than quantitative and difficult separations from side products may be required. Alternative methods may often be considered but they are frequently unselective, inefficient, and/or involve difficult, hazardous, or exotic fluorine reagents.

REFERENCES

1. M.A. McClinton and D. McClinton, *Tetrahedron,* 1992, 48, 6555.
2. K. Uneyama, *J. Synth. Org. Chem. Jpn.,* 1991, 49, 612.
3. D. Seebach, *Angew. Chem. Int. Ed. Engl.,* 1990, 29, 1320.
4. N. Muller, *J. Pharm. Sci.,* 1986, 75, 987.
5. V. Reiffenrath, J. Krause, H.L. Plach and G. Weber, *Liquid Crystl.,* 1989, 5, 159.
6. G.W. Gray, M. Hird and K.J. Toyne, *Mol. Crystl. Liq. Crystl.,* 1991, 206, 205.
7. H. Smith, 1992, D.Phil. thesis, University of York (U.K).
8. P.E. Cassidy, *Polym. Prep.,* 1990, 31, 338.
9. P.E. Cassidy, T.M. Aminabhavi and J.M. Farley, *J. Macromol. Sci. Rev. Macromol. Chem. Phys.,* 1989, C29, 365.
10. A.G. Farnham and R.N. Johnson, U.S. Patent 3,332,909 (1967).
11. T. Ichino, S. Sasaki, T. Matsura and S. Nishi, *J. Polym. Sci. Part A-1,* 1990, 28, 232.
12. T. Matsuura, N. Yamada, S. Nishi and J. Hasuda, *Macromolecules,* 1993, 26, 419.
13. W.R. Shiang and E.P. Woo, *J. Polym. Sci. Part A,* 1993, 31, 2081.
14. J.H. Clark and J.E. Denness, *Polymer,* 1994, 35, 5124.
15. Z.G. Gardlund, *Polymer,* 1993, 34, 1850.
16. G. Maier, R. Hecht, O. Nuyken, K. Burger and B. Helmreich, *Macromolecules,* 1993, 26, 2583.
17. T. Matsuura, Y. Hasuda, S. Nishi and N. Yamada, *Macromolecules,* 1991, 24, 5001.
18. J.H. Clark and D.K. Smith, *Tetrahedron Lett.,* 1985, 26, 2227.
19. K. Hasokawa, S. Fugii and K. Inukai, *Nippon Kagaku Kaishi,* 1980, 1304 (Chem. Abstr., 1981, 94, 45498a).
20. J.H. Simons and R.E. McArthur, *Ind. Eng. Chem.,* 1947, 39, 364.
21. H. Kimoto and L.A. Cohen, *J. Org. Chem.,* 1979, 44, 2902.
22. H. Kimoto and L.A. Cohen, *J. Org. Chem.,* 1980, 45, 3831.
23. L. Strekowski, R.L. Wydra, M.T. Legla, A. Czarry, D.B. Harden, S.E. Patterson, M.A. Battiste, and J.M. Coxan, *J. Org. Chem.,* 1990, 55, 4777.
24. A.L. Henne and M.S. Newman, *J. Am. Chem. Soc.,* 1938, 60, 1697.
25. J. Riera, J. Cartaner, J. Carilla and A. Robert, *Tetrahedron Lett.,* 1989, 30, 3825.
26. J. Castaner, J. Riera, J. Carilla, A. Robert, E. Molins and C. Miravitlles, *J. Org. Chem.,* 1991, 56, 103.
27. P. Buu-Hoi, Nguyen D. Xuong and Nguyen V. Bac., *Compt. Rend.,* 1963, 257, 3182.
28. H. Ogoshi, H. Mizushima, H. Toi and Y. Aoyama, *J. Org. Chem.,* 1986, 51, 2366.
29. G. Meazza and G. Zanardi, *J. Fluorine Chem.,* 1991, 55, 199.
30. M. Hudlicky, *Chemistry of Organic Fluorine Compounds,* Pergamon Press, New York, 1961.
31. F. Swarts, *Bull. Akad. R. Belg.,* 1898, 35, 375.
32. J.H. Simons and C.J. Lewis, *J. Amer. Chem. Soc.,* 1938, 60, 492.
33. M.M. Boudakian, in *Kirk-Othmer: Encyclopedia of Chemical Technology,* Vol. 10, 3rd edition, eds. M. Grayson and D. Eckroth, Wiley, New York, p.901, 1980.
34. M.M. Boudakian, *J. Fluorine Chem.,* 1987, 36, 293.
35. T. Nakagawa, U. Hiramatsu and T. Honda, U.S. Patent 1983, 4, 393, 257.
36. Y. Ohsaka, U. Hiramatsu and T. Honda, U.S. Patent 1983, 4,400,563.
37. A.E. Feiring, *J. Fluorine Chem.,* 1972, 10, 375.
38. A.E. Feiring, U.S. Patent 1977, 4,051,148.
39. A. Marhold and E. Klauke, *J. Fluorine Chem.,* 1981, 18, 281.
40. A.J. Bloodworth, K.J. Bowyer and J.C. Mitchell, *Tetrahedron Lett.,* 1987, 28, 5347.
41. V.C.R. McLoughlin and J. Thrower, U.S. Patent 1968, 3,408,411.
42. Y. Kobayasti, K. Yamanoto and I. Kumadaki, *Tetrahedron Lett.,* 1979, 42, 4071.
43. G.E. Carr, R.D. Chambers, T.F. Holmes and D.G. Parker, *J. Chem. Soc. Perkin Trans 1,* 1988, 921.
44. N.V. Kondratenko, E.P. Vechirko and L.M. Yagupol'skii, *Synthesis,* 1980, 932.
45. T. Umemoto and A. Ando, *Bull. Chem. Soc. Jpn.,* 1986, 59, 447.
46. J.M. Paratian, S. Sibille and J. Périchon, *J. Chem. Soc. Chem. Commun.,* 1992, 53.

47. D.M. Wiemers and D.J. Burton. *J. Am. Chem. Soc.*, 1986, 108, 832.

48. M.A. Willert-Porada, D.J. Burton and N.C. Baenziger, *J. Chem. Soc. Chem. Commun.*, 1989, 1633.

49. D.J. Burton, D.M. Wiemers and J.C. Easdon, U.S. Patent 1986, 4,582,921.

50. Q.-Y. Chen and S.-W. Wu, *J. Chem. Soc. Chem. Commun.*, 1989, 705.

51. Q.-Y. Chen and J.-X. Duan, *J. Chem. Soc. Chem. Commun.*, 1993, 1389.

52. D.-B. Su, J.-X. Duan and Q.-Y. Chen, *Tetrahedron Lett.*, 1991, 7689.

53. H. Urata and T. Fuchikami, *Tetrahedron Lett.*, 1991, 91.

54. J.H. Clark, M.A. McClinton and R.J. Blade, *J. Chem. Soc. Chem. Commun.*, 1988, 638.

55. J.H. Clark, M.A. McClinton, C.W. Jones, P. Landon, D. Bishop and R.J. Blade, *Tetrahedron Lett.*, 1989, 2133.

56. J.H. Clark, J.E. Denness, M.A. McClinton and A.J. Wynd, *J. Fluorine. Chem.*, 1990, 50, 411.

57. J.M. Birchall, G.P. Irvin and R.A. Bayson, *J. Chem. Soc. Perkin Trans. 2*, 1975, 435.

58. Y. Kobayashi, I. Kumadaki, A. Ohsawa, S.I. Murakami and T. Nakano, *Chem. Pharm. Bull.*, 1978, 26, 1247.

59. Y. Girard, J.-G. Atkinson, P.C. Bélanger, J.J. Fuentes, J. Rokach, C.S. Rooney, D.C. Remy and C.A. Hunt, *J. Org. Chem.*, 1983, 48, 3220.

60. M. Yashida, T. Yashida, M. Kobayashi and N.J. Kamigata, *J. Chem. Soc. Perkin Trans. 1*, 1989, 909.

61. E.S. Huyser and E. Bedard, *J. Org. Chem.*, 1964, 29, 1588.

62. A.B. Cowell and C. Tamborski, *J. Fluorine Chem.*, 1981, 17, 345.

63. J.H. Simons and E.-O. Ramler, *J. Am. Chem. Soc.*, 1943, 65, 389.

64. B.R. Langlois, E. Laurent and N. Roidot, *Tetrahedron Lett.*, 1991, 32, 7525.

65. C. Wakselman and M. Tordeux, *J. Chem. Soc. Chem. Commun.*, 1987, 1701.

66. M. Tordeux, B. Langlois and C. Wakselman, *J. Chem. Soc. Perkin Trans. 1*, 1990, 2293.

67. Y. Tanabe, N. Matsuo and N. Ohno, *J. Org. Chem.*, 1988, 53, 4582.

68. A. Gregoric and M. Zupan, *J. Org. Chem.*, 1979, 44, 4120.

69. D. Naumann and J. Kischkewitz, *J. Fluorine Chem.*, 1990, 47, 283.

70. D. Naumann, B. Wilkes and J. Kischkewitz, *J. Fluorine Chem.*, 1985, 30, 73.

71. H. Sawada, M. Nakayama, M. Yoshida, T. Yoshida and N. Kamigata, *J. Fluorine Chem.*, 1990, 46, 423.

72. C. Lai and T.E. Mallouk, *J. Chem. Soc. Chem. Commun.*, 1993, 1359.

73. D. Cech and B. Schwarz, *Nucleic Acids Symp. Ser.*, 1991, 9, 29 (*Chem. Abstr.*, 1981, 95, 220035s).

74. T. Umemoto and O. Miyano, *Tetrahedron Lett.*, 1982, 23, 3929.

75. L. Hein and D. Cech, *Z. Chem.*, 1977, 17, 415 (*Chem. Abstr.*, 1978, 88, 37747s).

76. L. Hein, D. Cech and C. Liebenthal, *Ger (East)*, 1977, 119, 427 (1977) (*Chem. Abstr.*, 1978, 86, 106645e).

77. H. Kimoto, S. Fujii and L.A. Cohen, *J. Org. Chem.*, 1984, 49, 1060.

78. H. Kimoto, S. Fujii and L.A. Cohen, *J. Org. Chem.*, 1982, 47, 2867.

79. M. Nishida, H. Kimoto, S. Fujii, H. Hayakawa and L.A. Cohen, *Bull. Chem. Soc. Jpn.*, 1991, 64, 2255.

80. J.H.D. Utley and R.J. Holman, *Electrochem. Acta*, 1976, 21, 987.

81. J.H. Tobin, U.S. Patent 1977, 4,038,331.

82. V.M. Labroo, R.B. Labroo and L.A. Cohen, *Tetrahedron Lett.*, 1990, 31, 5705.

83. T. Umemoto and S. Ishihara, *Tetrahedron Lett.*, 1990, 31, 3579.

84. T. Umemoto and S. Ishihara, *J. Am. Chem. Soc.*, 1993, 115, 2156.

85. W.R. Hasek, W.C. Smith and V.A. Engelhardt, *J. Am. Chem. Soc.*, 1960, 82, 543.

86. W. J. Middleton, U.S. Patent 1975, 3,914,265 and U.S. Patent 1976, 3976691.

87. L.C. Rinzema, J. Stoffelsma and J. Arens, *Rect. Trav. Chim.*, 1959, 78, 354.

88. R. Breslow and D.S. Pandey, *J. Org. Chem.*, 1980, 45, 740.

89. D.P. Matthews, J.P. Whitten and J.R. McCarthy, *Tetrahedron Lett.*, 1986, 27, 4861.

90. Houben-Weyl, *Methoden der Organischem Chemie*, Band IX, ed. E. Mueller, Georg Thieme Verlag, Stuttgart, 1955.

91. M. Kuroboshi and T. Hiyama, *Chem. Lett.*, 1992, 827.

92. S. Rozen and E. Mishani, *J. Chem. Soc. Chem. Commun.*, 1994, 2081.

93. C.W. Jones, J.P. Sankey and W.R. Sanderson, *Fluorine in Agriculture*, Fluorine Technology, Cheshire, 1994.

Chapter 6

Trifluoromethylthioaromatics and Trifluoromethylsulfonylaromatics

6.1 INTRODUCTION

The combination of very high lipophilicity and high electron-withdrawing ability for the trifluoromethylthio group and the extremely high electron-withdrawing ability of the trifluoromethylsulfonyl group, along with their good physical and chemical stabilities, makes these groups extremely interesting substituents in aromatic chemistry. While applications for trifluoromethylthioaromatics and trifluoromethylsulfonylaromatics have been known for some 40 years,[1-3] the accessibility was restricted by the limited number of established routes, often employing harsh conditions, to these molecules. In more recent years, a number of new routes, notably the development of nucleophilic trifluoromethylthio reagents, have made these molecules more accessible and interest in their preparation, properties, and applications is increasing.

The oldest applications for trifluoromethylthioaromatics and trifluoromethylsulfonylaromatics are probably in the dyes industry.[1-3] More recently, applications have been extended to pharmaceuticals, agrochemicals, and, in the case of trifluoromethylthioaromatics, liquid crystals.[3,4] Trifluoromethylthioaromatic isocyanates are useful intermediates for biologically active compounds, including herbicides and insecticides. The dichlorofluorothio and chlorodifluorothio substituents have also proven to be popular in this context due to the lower cost of production. Thus, compound **I** has at least the same range of activity as compound **II** for weed control in sown and transplanted rice, but **II** is favored on cost grounds. This can be traced back to the traditional method of synthesis of trifluoromethylthioaromatics by acid-catalyzed halogen exchange of the corresponding trichloromethylthioaromatics (see Section 6.4.4). The newer-generation, commercial trifluoromethylthioaromatic compounds can be made by other methods. Thus, the very promising pesticide Fipronil™,[5] is now manufactured under reductive conditions using $CF_3Br/Na_2S_2O_4$.[6]

6.2 PHYSICAL PROPERTIES OF TRIFLUOROMETHYLTHIOAROMATICS AND TRIFLUOROMETHYLSULFONYLAROMATICS

The trifluoromethylthio group is in many ways similar to the trifluoromethoxy functionality, although the sulfur is larger and more polarizable than oxygen. The presence of the electronegative trifluoromethyl group results in a contraction of the d orbitals on sulfur through increasing the overlap between the trifluoromethylthio moiety and the adjacent pi system.[7] The trifluoromethylthio group is certainly larger

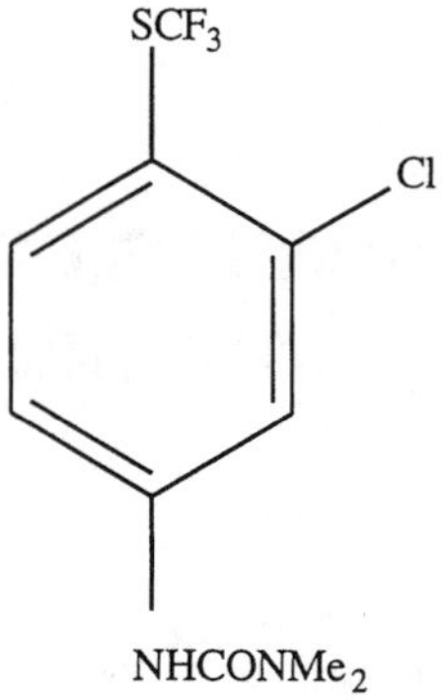

I

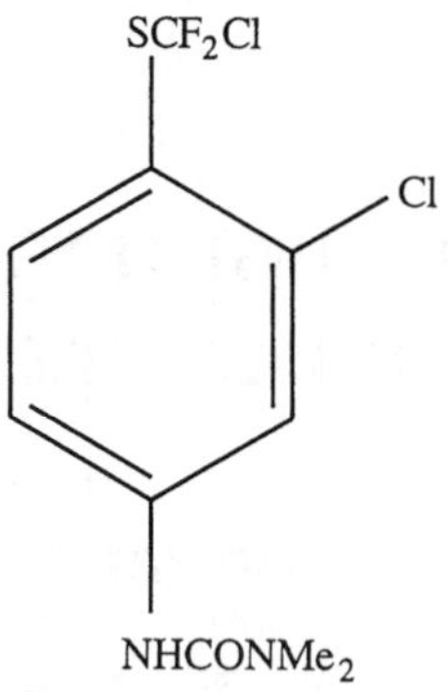

II

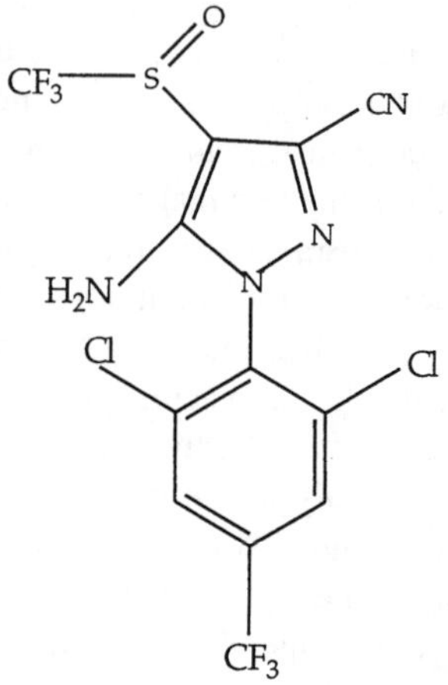

Fiprunoil (insecticide)

than the trifluoromethyl group, and this is particularly important in the context of liquid crystals, since the size of the substituent affects the way the molecules pack in the crystalline and subsequent mesophases.[8] This means that the phases exhibited by the molecules containing the trifluoromethyl group (a very popular substituent in liquid phases) are thermally more stable. Trifluoromethylthioaromatic liquid crystals can also show lower birefringence values compared to other fluorinated analogues and this can also be attributed to the bulkiness of the substituent increasing the volume of the molecule.[9]

The lipophilicity induced in aromatic molecules by the trifluoromethylthio substituent is probably the most important of its effects. The lipophilicity parameter for this group is significantly greater than that for trifluoromethoxy, trifluoromethyl, trichloromethylthio, and indeed most small aromatic substituents (Table 6.1).[10]

Table 6.1 Lipophilic Parameters for Monosubstituted Benzenes (Measured as the 1-octanol-Water Partition Coefficient)

PhH	ca.1.8[a]
PhCH$_3$	ca.2.5[a]
PhF	2.27
PhSO$_2$CF$_3$	2.27
PhCF$_3$	2.79
PhCl	2.84
PhCCl$_3$	2.92
PhOCF$_3$	3.17
PhSCF$_3$	3.79

[a] Mean of several values.

Measured substituent parameters for the trifluoromethylthio group in aromatic molecules suggest that the group is generally electron withdrawing, as might be expected. The direct inductive electron-withdrawing ability is smaller than, or at best comparable to, that of F, CF$_3$, OCF$_3$, and even Cl, but it is clearly greater than that of SCH$_3$ or OCH$_3$[7] (Table 6.2). The ability of the group to stabilize negative charge through resonance effects can be thought of in terms of the three valence bond structural descriptions:

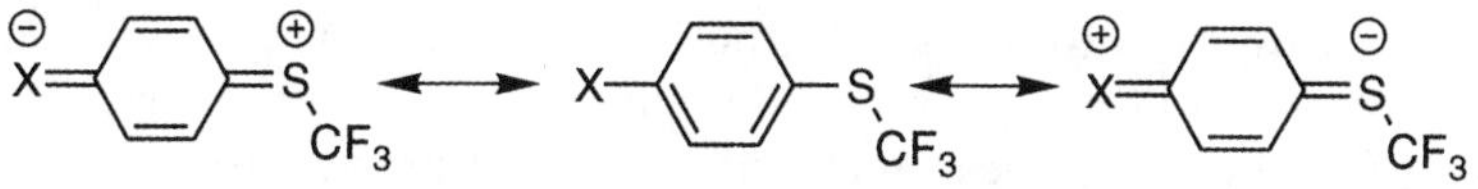

While OCF$_3$ and SCH$_3$, as well as F, show significant contributions from structures similar to **I** (i.e., destabilizing negative charge in the ring), SCF$_3$ appears to encourage large contributions from **III**.[7]

Table 6.2 Substituent Parameters

SO$_2$CF$_3$	0.79	0.48
SCF$_3$	0.37	0.18
OCF$_3$	0.47	−0.17
CF$_3$	0.39	0.11
F	0.45	−0.40
SO$_2$CH$_3$	0.62	0.16
SCH$_3$	0.22	−0.28
OCH$_3$	0.21	−0.47
Cl	0.42	−0.25

The trifluoromethylsulfonyl group is the strongest neutral electron-withdrawing group. Like the nitro group, its electron-withdrawing power is a result of significant resonance interaction as well as the inductive mechanism. Generally, the acidities of substituted phenols, for example, increase in the order of substituents, H < OCF$_3$ < SCF$_3$ < SO$_2$CF$_3$.[7] Generally, the replacement of aliphatic hydrogens by fluorine causes a decrease in boiling point and this is easily witnessed in the series $PhSCF_3$ (b.p. 142°C) through to $PhSCH_3$ (b.p. 190°C).

The thermal stability of trifluoromethyl thioaromatics is poorer than that of the corresponding ethers,[11] with the former decomposing in a sealed tube at about 500°C, whereas the latter survive to almost 600°C.

6.3 CHEMICAL PROPERTIES OF TRIFLUOROMETHYLTHIOAROMATICS AND TRIFLUOROMETHYLSULFONYLAROMATICS

The trifluoromethylthio group is chemically stable to a number of standard reaction conditions, enabling the preparation of a wide variety of aromatic molecules containing the group. Thus, trifluoromethylaromatics can be treated with halogens, strong acids (enabling ring halogenation and nitration) and with hydrogenation reagents (e.g., Pt/H_2). Electrophilic substitution reactions generally occur predominately at the para position (Figure 6.1). Strong oxidizing agents will, however, convert the trifluoromethylthio group to the trifluoromethylsulfonyl group (Figure 6.1).

The trifluoromethylsulfonyl group is potentially vulnerable to substitution on attack by a nucleophile. In competitive systems, it can be more resistant than a comparably activated nitro or chlorine group (Figure 6.2). Nonetheless, nucleophilic substitution reactions on ring-substituted trifluoromethylsulfones is likely to lead to reduced efficiency due to a loss in the trifluoromethylsulfonyl group.[12]

6.4 SYNTHESIS OF TRIFLUOROMETHYLTHIOAROMATICS AND TRIFLUOROMETHYLSULFONYLAROMATICS

As with the preparation of trifluoromethylaromatics, trifluoromethylthioaromatics have traditionally been prepared by halogen exchange of the intermediate trichloro compounds. While such methods are well established, especially in industry, the harsh methods required are a serious drawback in terms of health and safety, plant life-time, and possible acid-catalyzed side reactions. The lower electron-withdrawing power of the trifluoromethylthio group compared to the trifluoromethyl group makes the design and use of electrophilic reagents more realistic and methods based on these are quite well established. Radical reactions en route to trifluoromethylthioaromatics are also well known, although these commonly occur via radical-anion mechanisms. As with trifluoromethylation, the nucleophilic substitution of the trifluoromethylthio group is a relative newcomer on the scene that has quickly developed into a versatile and relatively simple technology. Trifluoromethylthiocopper, $CuSCF_3$, is already being used as a source of SCF_3 in the agrochemicals industry.

The direct incorporation of the trifluoromethylsulfonyl group into an aromatic substrate is largely limited to the use of electrophilic reagents, often in tandem with a metal aryl. Alternative methods of synthesis of trifluoromethylsulfonylaromatics do exist, of which the oxidation of trifluoromethylthioaromatics is one.

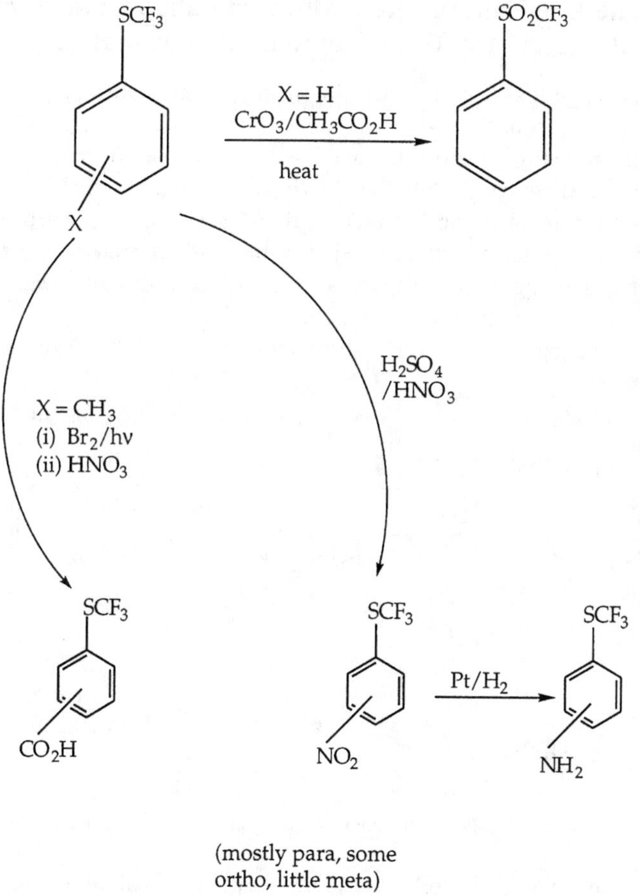

FIGURE 6.1 Reactions of trifluoromethylthioaromatics.

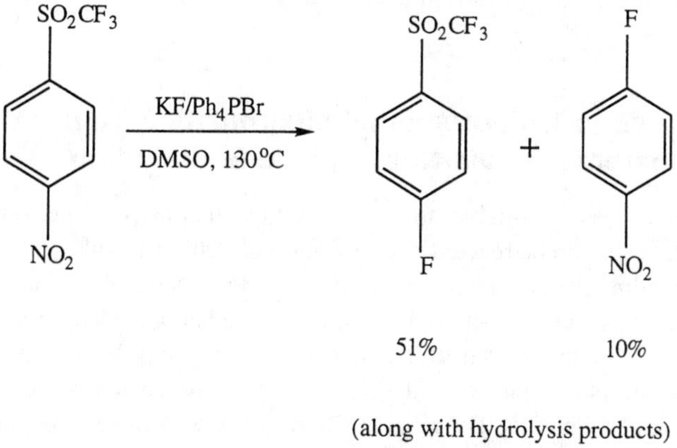

FIGURE 6.2 Competitive substitution by fluorine in aromatic trifluoromethylsulfones.

6.4.1 Routes to Trifluoromethylthioaromatics from Arylmethyl Sulfides via the Trichloromethylthioaromatics

Halogen exchange fluorination of trichloromethylthioaromatics prepared by photo-initiated chlorination of the corresponding arylmethyl sulfides is a well-established route to trifluoromethylthioaromatics and is essentially based on traditionally Swarts-type chemistry (Figure 6.3).[13-18] The fluorination reaction, which can proceed in reasonable yield, requires the Lewis acid, SbF_3, and is reminiscent of the halogen exchange reaction route to trifluoromethylaromatics. This route is effective for aromatic rings containing inert substituents, such as chloro and nitro, but more severe conditions, along with the use of catalytic quantities of boron trifluoride, may be required in some cases[18] and ortho substituents can render the route ineffective.[9]

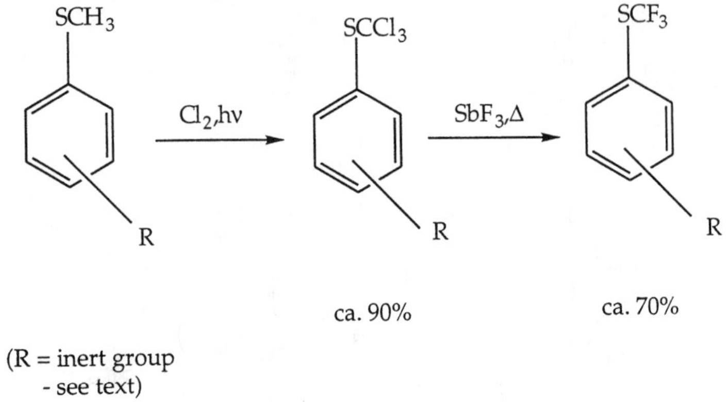

ca. 90%   ca. 70%

(R = inert group
  - see text)

FIGURE 6.3 Synthesis of trifluoromethylthioaromatics from the corresponding aryl methylsulfides.

Trifluoromethylthiophenothiazines have pharmaceutical utility and can be synthesized by using the SbF_3 reaction in the first, key fluorination step (Figure 6.4).[19] It is interesting to note that the fluorinating agent, being a Lewis acid and hence electrophilic, does not substitute the ring fluorine (which would require a nucleophilic reagent).

6.4.2 Routes to Trifluoromethylthioaromatics from Haloaromatics via Nucleophilic Substitution

Halogen derivatives of suitable aromatic systems such as pyridine, quinoline, and activated benzenes can be reacted with trifluoromethylthio-metal compounds to give the corresponding trifluoromethylthioaromatics.[9,18,20] Several trifluoromethylthio-metal compounds are known to be useful in aliphatic nucleophilic substitution chemistry such as the mercury (II) and silver (I) compounds, but nucleophilic aromatic substitution of aromatic halogen is normally carried out using trifluoro-methylthiocopper, $CuSCF_3$. The reagent can be prepared from copper metal by three main methods (Figure 6.5).[9,21] It decomposes in the presence of oxygen and a solvent or a support material (Figure 6.6).[22] Stability is relatively high in acetonitrile,

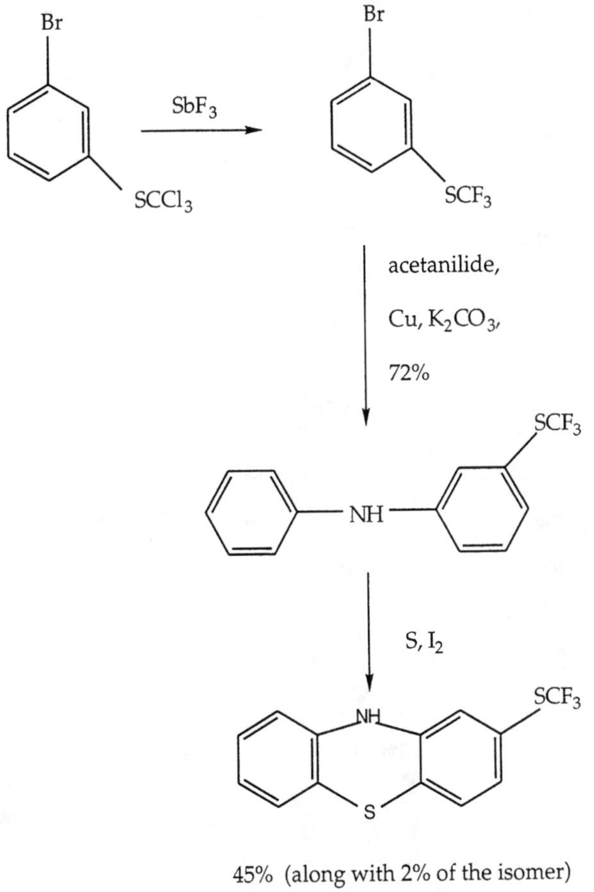

FIGURE 6.4   Route to trifluorothiomethylated phenothiazines.

dimethylformamide, and dimethylsulfoxide, but relatively low in pyridine and hexa-methylphosphoramide. Supported $CuSCF_3$ can be a more active form of the reagent but it is vulnerable to decomposition in the presence of oxygen (Figure 6.7).

Trifluoromethylthiocopper will react with iodoheteroaromatics to give the corresponding trifluoromethylthioheterocyclics (Figure 6.8). Iodoaromatics and, to a lesser extent, bromoaromatics will also react with the reagent — the presence of electron-withdrawing groups generally leads to good product yields (70 to 75%), whereas electron-donating substituents lead to relatively poor yields (30 to 55%).[18] The presence of ortho substituents is not detrimental[21] (Figure 6.9). Multiple substitution is possible and even the fully substituted hexa(trifluoromethylthio)benzene can be prepared by this method (Figure 6.10).[23]

A convenient alternative to preformed $CuSCF_3$ is to prepare the reagent *in situ*. This can be accomplished by the reaction of $FSO_2CF_2CO_2Me$ with Cu (I) in the presence of S_8 in a dipolar aprotic solvent such as HMPA or NMP (interestingly DMF is not effective) (Figure 6.11).[24]

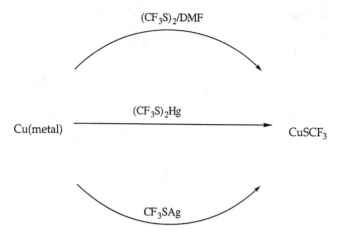

FIGURE 6.5   Routes to trifluoromethylthiocopper.

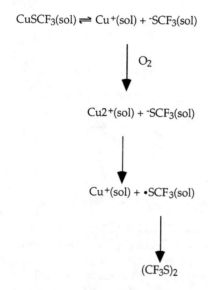

FIGURE 6.6   Decomposition of CuSCF$_3$; sol, suitable solvent (HMPA, pyridine) or support material.

Non-copper-based trifluoromethylthio-metal compounds may be capable of reacting with highly activated aromatic substrates such as perfluoropyridines,[25,26] perfluoropyrazine,[27] and perfluoropyrimidine[27] (Figure 6.12). Here the reagent is generated *in situ* from the reaction of thiocarbonyl fluoride, F$_2$C=S, and potassium fluoride[28] or bis(trifluoromethyl)thiocarbonate, (CF$_3$)$_2$C=S and a fluoride ion source.[26] The former gives the better yield, but the latter is more practically convenient, since it involves a liquid rather than a gas.

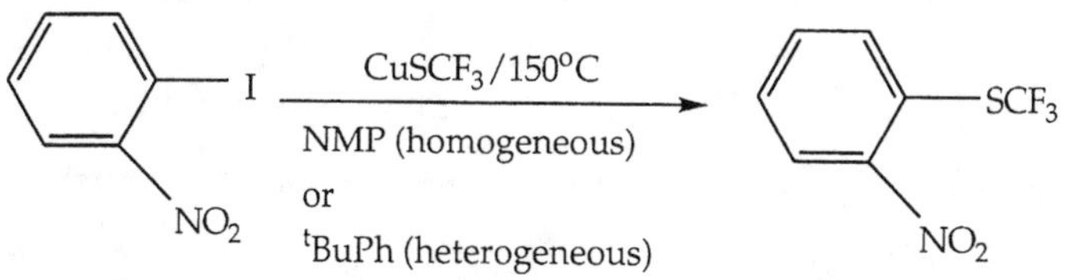

Support	Yield/% (time/h)
—	100 (2)
Alumina	100 (2)
Charcoal	40 (8)
Silica Gel	0 (24)

FIGURE 6.7 Relative reactivities of CuSCF$_3$ and supported CuSCF$_3$.

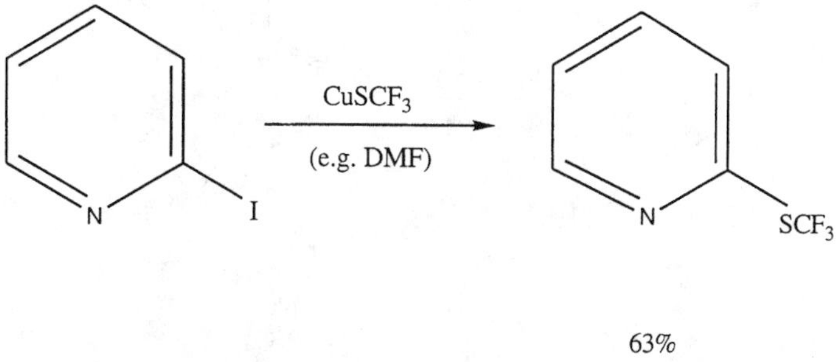

63%

FIGURE 6.8 Reaction of CuSCF$_3$ with an iodopyridine.

6.4.3 Routes to Trifluoromethylthioaromatics from Aromatic Hydrocarbons via Electrophilic Substitution

Unlike the trifluoromethyl group, the less electronegative trifluoromethylthio group can be readily incorporated into aromatic substrates by electrophilic substitution. Trifluoromethylsulfenyl chloride, CF$_3$SCl (which can be prepared by reaction of CF$_3$SSCF$_3$ on (CF$_3$S)$_2$Hg with chlorine[29]) is commonly used as the source of CF$_3$S in electrophilic aromatic substitutions. Highly activated aromatic substrates such as N,N-dimethylaniline and phenol will react directly with the reagent to give almost exclusive parasubstitution (Figure 6.13).[30] Less active substrates, such as benzene and the halobenzenes (which give a mixture of isomers) only undergo reaction on heating and in the presence of a catalyst. The reaction can be considered as a Friedel Crafts reaction and the suitable catalysts are powerful Brönsted acids, such as hydrogen fluoride, or powerful Lewis acids, such as boron trifluoride (Figure 6.13).[30] Polysubstitution is not a problem, as the CF$_3$S group is highly deactivating towards

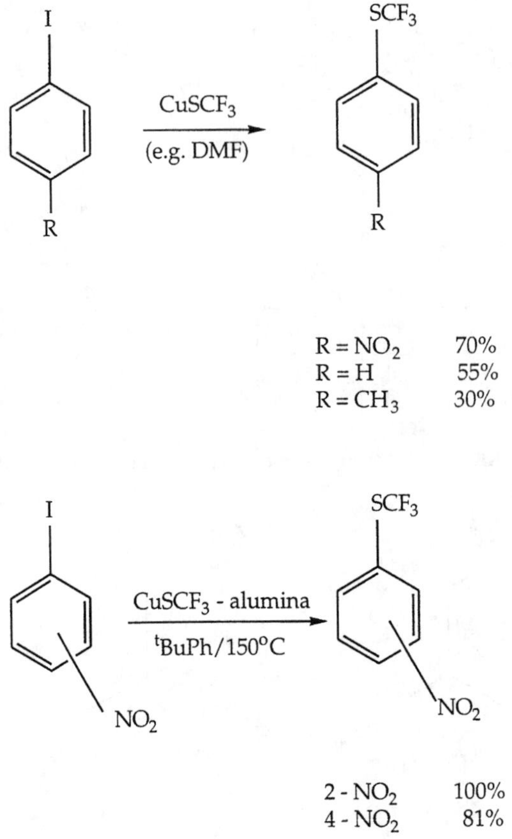

$$R = NO_2 \quad 70\%$$
$$R = H \quad\quad 55\%$$
$$R = CH_3 \quad 30\%$$

$$2 - NO_2 \quad 100\%$$
$$4 - NO_2 \quad\; 81\%$$

FIGURE 6.9   Reactions of CuSCF$_3$ with iodobenzenes.

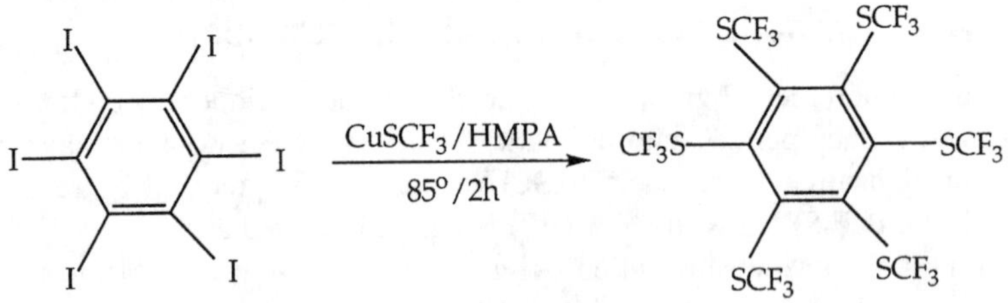

FIGURE 6.10   Preparation of hexa(trifluoromethylthio)benzene.

further electrophilic substitution. Interestingly, thiophenol only reacts to give sub-stitution at sulfur (giving PhSSCF$_3$) and it may be that the reaction of phenol involves initial formation of the sulfenate ester, ArOSCF$_3$, followed by rearrangement to give the observed product.

Activated heteroaromatics, such as furan, pyrrole, and imidazole, will also react directly with trifluoromethylsulfenyl chloride.[9] Disubstitution may occur if excess

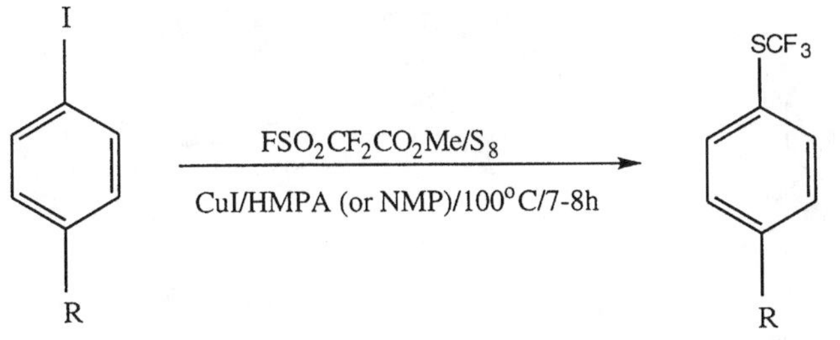

$$R = H \quad\quad 60\%$$
$$R = Cl \quad\quad 0\% \quad\quad (43\% \text{ in NMP})$$
$$R = NO_2 \quad 63\%$$
$$R = Me \quad\quad 41\% \quad\quad (NMP)$$

FIGURE 6.11 *In situ* generation and reaction of $CuSCF_3$.

chloride is used.[9] Pyridine is not reactive enough and will only react as an organo-metallic reagent (e.g., as an aluminate — Figure 6.14). Generally less active aromatic substrates can be made active towards attack by trifluoromethylsulfenyl chloride by converting them into organometallic reagents (typically Grignards — Figure 6.14). This can be considered as being an alternative strategy to the use of Friedel Crafts conditions, as described above. Unfortunately the use of Grignard reagents can lead to mixed products and low reaction efficiencies. Low-temperature reactions, in particular, can lead to simultaneous formation of the aryl halides (the halide anion X coming from the Grignard ArMgX as well as from the trifluoromethylsulfenyl chloride, (X = Cl)). It is often advisable to accept a lower yield but to ensure separation by avoiding X = Br, which leads to a bromoaromatic product that may often have a boiling point comparable to the trifluoromethylthioaromatic.[31]

6.4.4 Routes to Trifluoromethylthioaromatics from Aromatic Thiols via Radical Substitution

Simple radical substitution routes to trifluoromethylthio aromatics are unusual. Pentafluoroiodobenzene reacts with bis(trifluoromethyl)disulfide to give a mixture of products that suggests that a pentafluorophenyl radical is produced which attacks the disulfide, displacing a trifluoromethyl radical, which subsequently abstracts iodine (Figure 6.15).[32]

The reactions of thiols and their salts with a source of the trifluoromethyl group is synthetically useful. Highly activated thiols and pyrimidines will react directly with trifluoromethyl iodide under photochemical activation (Figure 6.16).[33] A more useful reaction is that of the potassium salts of thiols with trifluoromethyl halides (bromide or iodide), which is quite general in giving the trifluoromethylthioaromatic (Figure 6.16).[34] Yields are variable (7 to 75%) and electron-donating substituents do enhance the product yields. Reactions are thought to proceed via a radical-anion mechanism (Figure 6.17). An alternative route is to start from an aromatic disulfide in the presence of a sulfur dioxide radical anion generated from sodium dithionate,

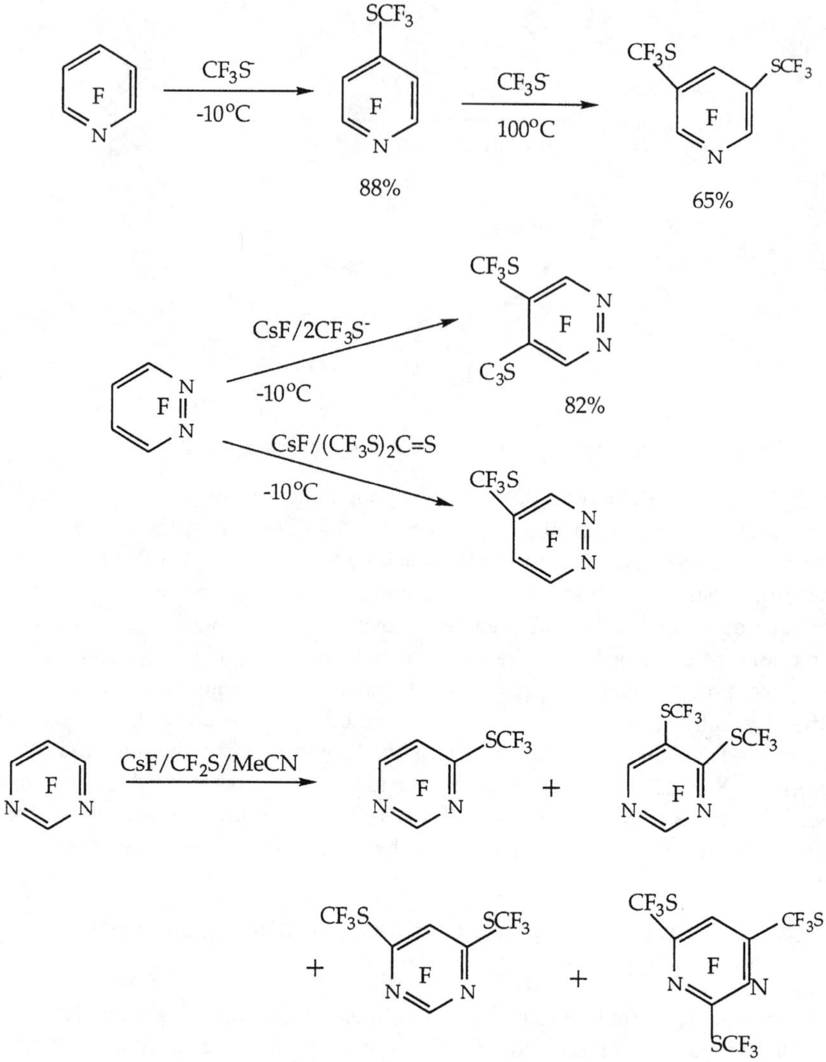

FIGURE 6.12   Reactions of perfluoroheteroaromatics with CF_3S^-.

$Na_2S_2O_4$, or sodium hydroxymethane sulfinate, NaO_2SCH_2OH.[35] Reactions based on the readily available and relatively inexpensive trifluoromethyl bromide have real commercial potential.

6.4.5   Routes to Trifluoromethylsulfonylaromatics from Aromatic Hydrocarbons via Electrophilic Substitution

By analogy with the reactions of the electrophilic reagent CF_3SCl with aryl metal reagents, it might be expected that a similar approach can be used to make trifluoromethylsulfonyl aromatics (see Section 6.4.4). Reaction of phenyl lithium with the

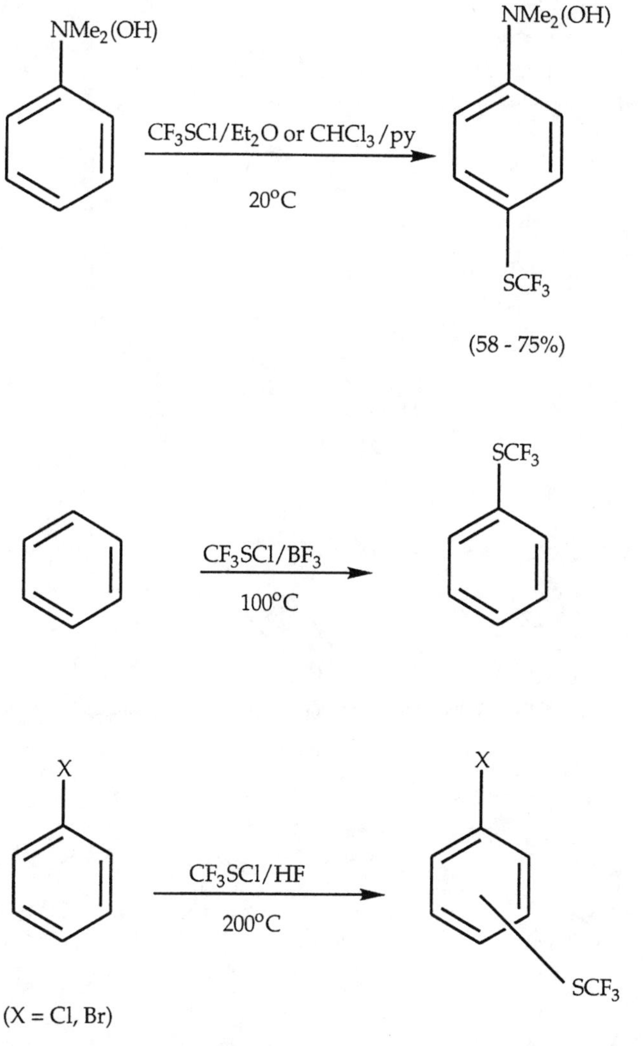

FIGURE 6.13 Direct trifluoromethylthiolation of aromatic substrates.

electrophilic reagent trifluoromethylsulfonic anhydride, $(CF_3SO_2)_2O$, in fact gives the diphenylsulfone as the final product (Figure 6.18).[36] Successful electrophilic substitution requires Friedel Crafts conditions, typically the use of aluminum chloride, although this is a limited route.[37] Highly deactivated substrates such as nitrobenzene and 1,4-dibromobenzene do not react with the weak acylating complex $(CF_3SO_2)_2O/AlCl_3$, while highly reactive substrates such as anisole produce a highly stable, colored complex which does not dissociate to the desired product. Substrates of intermediate activity — benzene, alkylbenzenes, and monohalobenzenes — can be successfully reacted to give the trifluoromethylsulfonylaromatic products (Table 6.3). Interestingly, the more expected reagent, trifluoromethylsulfonyl chloride, is unreactive towards aromatic substrates.[37]

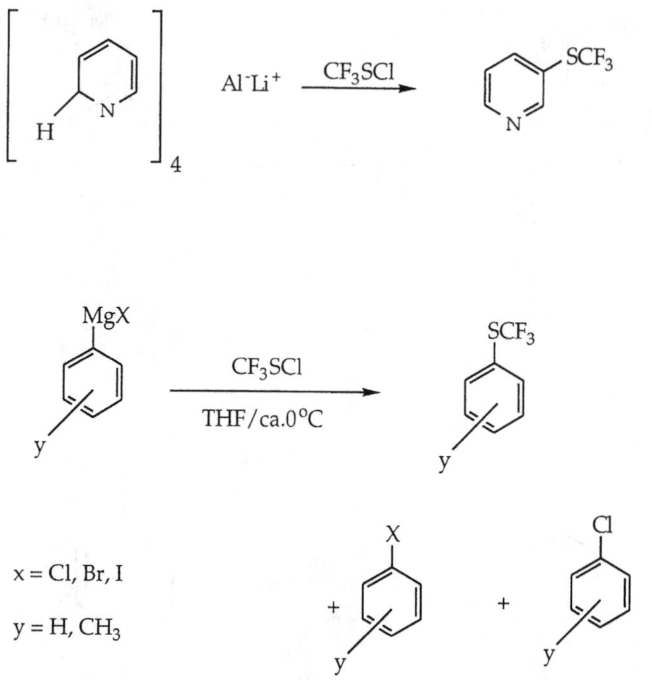

FIGURE 6.14 Trifluoromethylthiolation of aromatic organometallic reagents.

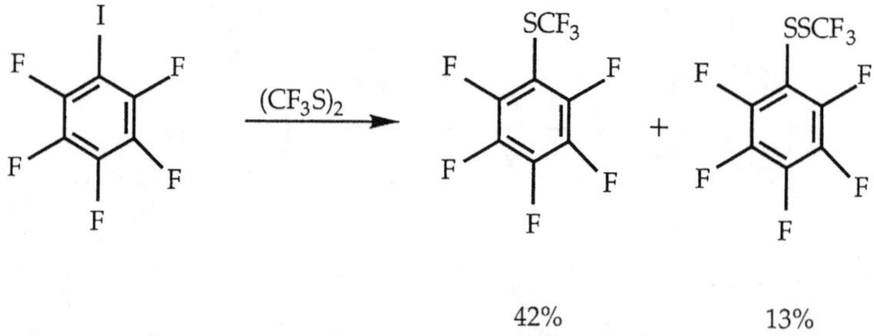

42%   13%

FIGURE 6.15 Radical reactions of pentafluoroiodobenzene and bis(trifluoromethyl)sulfide.

6.4.6 Routes to Trifluoromethylsulfonylaromatics from Trifluoromethylthioaromatics via Oxidation

Trifluoromethylthioaromatics can generally be oxidized to the corresponding sulfone[9,12,37] using chromium trioxide in ethanoic acid, although milder conditions such as hydrogen peroxide may be sufficient.[24] Yields are variable, however, and can be low in the case of 2-substituted aromatic substrates (Table 6.4), and care must be taken to avoid further reaction resulting from the strong activating power of the

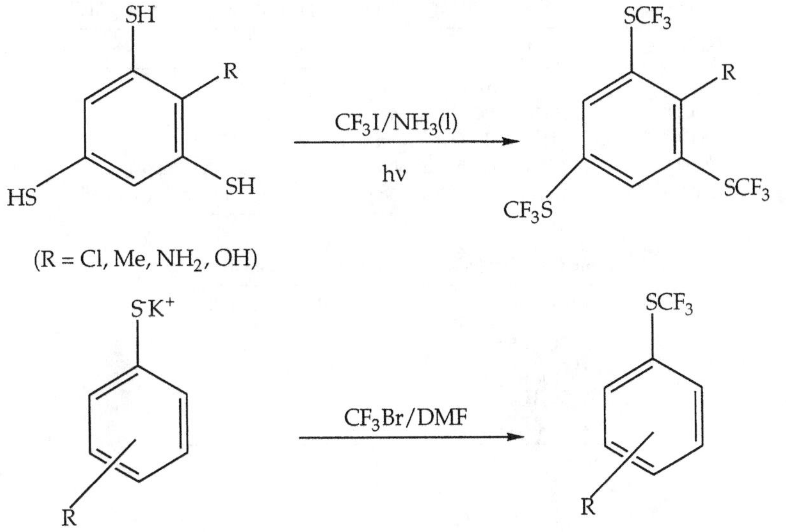

$$(R = Cl, Me, NH_2, OH)$$

R = H, 4-Me, 3-MeO,   7 - 75%
2-MeO, 3-NH$_2$, 3CF$_3$, 4-Cl

FIGURE 6.16   Reactions of thiols and thiolates with trifluoromethyl halides.

FIGURE 6.17   Possible mechanism for the reaction of thiophenoxide with trifluoromethyl-bromide.

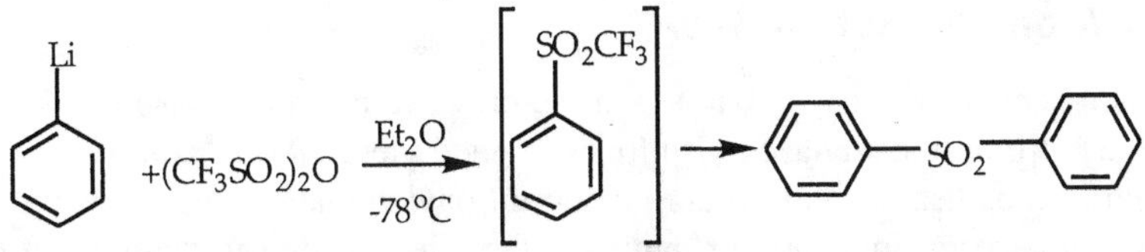

$$ArS^- + CF_3Br \longrightarrow ArS\bullet + [CF_3Br]^{\bullet -}$$

$$[CF_3Br]^{\bullet -} \longrightarrow CF_3\bullet + Br^-$$

$$ArS^- + CF_3\bullet \longrightarrow [ArSCF_3]^{\bullet -}$$

$$[ArSCF_3]^{\bullet -} + CF_3Br \longrightarrow ArSCF_3 + [CF_3Br]^{\bullet -}$$

FIGURE 6.18   Reaction of phenyllithium with (CF$_3$SO$_2$)$_2$O.

trifluoromethylsulfonyl group. Oxidation of trifluoromethylthio-substituted heteroaromatics to the corresponding sulfone can also be achieved in this way, although ring hydroxylation can also occur when using aqueous mineral acid solvents (Figure 6.19).[38] The major drawback of this route to trifluoromethylsulfonylaromatics is the need to synthesize the sulfide starting material, which may not be straightforward.

Table 6.3 Trifluoromethylsulfonylation of Aromatics Using $(CF_3SO_2)_2/AlCl_3$

Substrate	Product
PhH	$PhSO_2CF_3$ (61%)
$PhCH_3$	$MeC_6H_4SO_2CF_3$ (69%; mix of isomers)
PhCl	$ClC_6H_4SO_2CF_3$ (10%; mix of isomers)
$PhNO_2$	No trifluoromethylsulfone product; substrate decomposes
PhOMe	No trifluoromethylsulfone product; strong pi-complex formed

Table 6.4 Oxidation of Trifluoromethylthioaromatics

Substrate	Yield of Sulfone/%
$2\text{-}CF_3SC_6H_4NO_2$	17
$4\text{-}CF_3SC_6H_4NO_2$	64
$2\text{-}CF_3SC_6H_4Cl$	40
$4\text{-}CF_3SC_6H_4Cl$	75

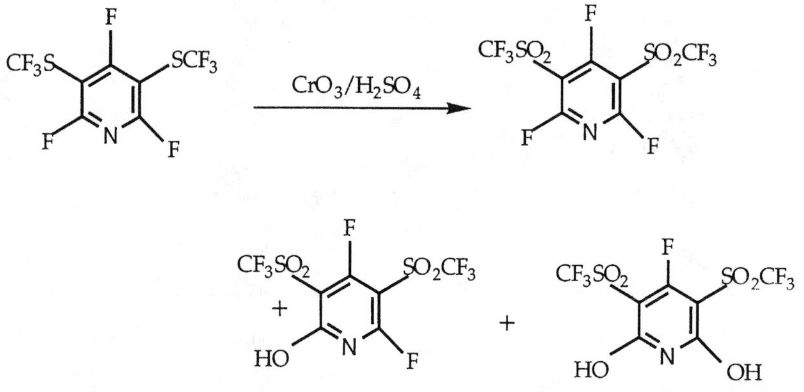

FIGURE 6.19 Oxidation of trifluoromethylthio-substituted perfluoropyridine.

6.4.7 Routes to Trifluoromethylsulfonylaromatics from other Aromatic Substrates

Aryl sulfonamides react with CF_3NO in the presence of base to give the thermally unstable N-trifluoromethylazoarylsulfones in good yields. These compounds eliminate nitrogen on heating and trifluoromethylsulfonylaromatics are formed by resulting radical cage recombination (Figure 6.20).[39] The reaction is reminiscent of the commercial route to fluoroaromatics based on thermal decomposition of aryl diazonium fluorides (see Chapter 3).

While it has proved possible to synthesize pentafluoroethylsulfonylaromatics by reaction of arylsulfonyl fluorides with tetrafluoroethene in the presence of F−,[40] this is not a viable route to the trifluoromethylsulfonularomatics due to the instability of CF_3^- even at −100°C (Figure 6.21).[41] An alternative approach is to use a source of difluorocarbene that can insert into the S−F bond of the arylsulfonyl fluoride. This

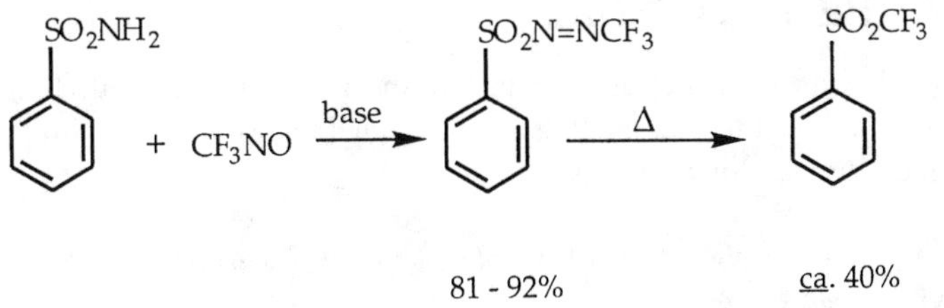

FIGURE 6.20 Formation and thermal decomposition of *N*-trifluoromethylphenylsulfone.

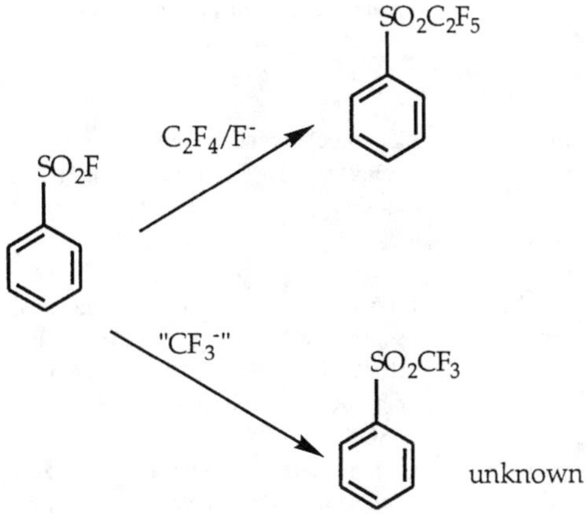

FIGURE 6.21 Possible reactions of perfluoroalkyl anions with benzene sulfonyl fluoride.

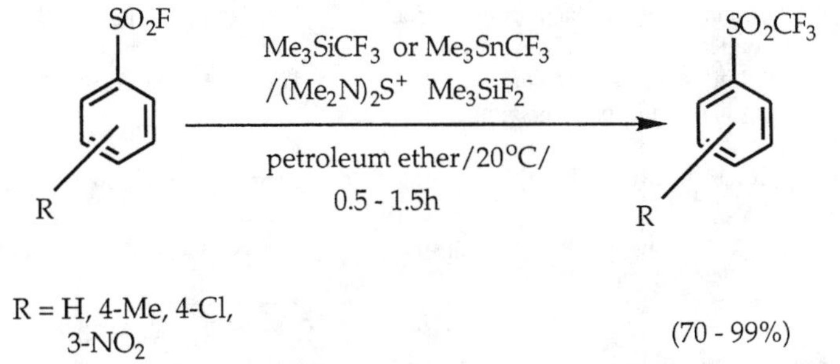

$R = H, 4\text{-Me}, 4\text{-Cl},$
3-NO_2

$(70 - 99\%)$

FIGURE 6.22 Reaction of difluorocarbene sources with arylsulfonyl fluorides.

can be accomplished by using (trifluoromethyl)trimethylsilane or the stannane analogue in the presence of a nucleophilic reagent, preferably $(Me_2N)_3S^+Me_3SiF_2^-$ (TASF) (Figure 6.22). Other substituents on the ring have little effect on the high reaction efficiency. The silane reagent operates effectively in polar and nonpolar solvents ranging from dimethylformamide to petroleum ether, whereas the stannane is more effective in the less polar solvents. Clearly the reaction is effective and widely applicable, although the cost of the reagents might be prohibitive.

6.5 SYNTHETIC METHODS

The major fluorine reagents used in the synthesis of trifluoromethylthioaromatics and trifluoromethylsulfonylaromatics are given in Table 6.5, along with guidance on their availability, handling, and use.

Table 6.5 Major Fluorine Reagents Used in the Synthesis of Trifluoromethylthioaromatics and Trifluoromethylsulfonylaromatics

Reagent	Availability/Preparation	Comments on Handling and Use
HF	Commercially available as anhydrous acid, aqueous solutions, pyridine complexes, and as polyfluoride	Anhydrous HF is a strong Brönsted acid and may cause acid-catalyzed reactions; strong solutions are very corrosive and must be handled with extreme care; HF attacks glass and should be handled in suitable plasticware
SbF_3 and other antimony halides	Commercially available as crystalline solids or prepared by action of HF on oxide	Solid sublimes at >300°C; highly water soluble; Sb(V) halides are all extremely powerful Lewis acids, forming strong complexes with Lewis bases including sulfides and sulfones
$CuSCF_3$/ supported $CuSCF_3$	Prepared by reaction of copper metal with sources of CF_3S, e.g., $(CF_3S)_2$/DMF; $(CF_3S)_2Hg$; CF_3SAg, or generate in situ, e.g., by reaction of Cu (I) with $FSO_2CF_2CO_2Me/S_8$	Can be isolated as a solid; supported $CuSCF_3$ and $CuSCF_3$ in strongly coordinating solvents (py, HMPA) can decompose in air
CF_3SCl	Commercially available from specialist outlets but very expensive and can be prepared by reaction of $(CF_3S)_2$ or $(CF_3S)_2Hg$ with chlorine or by halex of CCl_3SCl with NaF	CF_3SCl is a toxic gas (b.p. −3°C). $(CF_3S)_2Hg$ is highly toxic
$(CF_3S)_2$	Commercially available from specialist outlets but extremely expensive	Low boiling liquid (b.p. 34°C); toxic
CF_3X	All of the halides are commercially available; the iodide is significantly more expensive than the others	The halides are gases
$(CF_3SO_2)_2O$	Commercially available	Corrosive liquid; slowly reacts with cold water
CF_3NO	Commercially available from specialist outlets but very expensive	Low boiling gas (b.p. −85°C); highly toxic.

Antimony trifluoride is generally the most appropriate reagent for the conversion of $-SCCl_3$ to $-SCF_3$, although its activity may not always be sufficient. In these cases a more powerful Lewis acid catalyst such as BF_3 can be added to boost activity. The method may be ineffective if orthosubstituents are present and the method is perhaps less generally useful than it is for Ar-$CCl_3 \rightarrow ArCF_3$ conversions (Chapter 5).

In contrast to "$CuCF_3$", which has never been isolated, $CuSCF_3$ is a stable solid. This valuable property of the reagent is partly offset by the harsh conditions and toxic reagents which have traditionally been associated with its preparation. The best method of reagent preparation would seem to be the reaction of trifluoromethylthiosilver, $AgSCF_3$, and copper (I) bromide. The $AgSCF_3$ is prepared by reaction of silver (I) fluoride with carbon disulfide in a dipolar aprotic solvent.[21] As well as the reagent itself, the corresponding supported reagent, $CuSCF_3$-alumina (1 mmol $CuSCF_3$ g^{-1}) is a useful source of nucleophilic SCF_3. The latter is made by slow evaporation from a slurry of the alumina in an acetonitrile solution of the reagent. Both the reagent and the supported reagent are subject to decomposition. The reagent is unstable in solvents that encourage dissociation to Cu^+ (pyridine, HMPA) and in the presence of O_2. In the case of the supported reagent, the support itself seems to act as the "solvent" and decomposition occurs on heating in air.[22]

An apparent *in situ* route to $CuSCF_3$ is the reaction of the difluorocarbene source, $FSO_2CF_2CO_2Me$ with Cu (I) and sulfur:

$$FSO_2CF_2CO_2Cu \rightarrow \text{"CuF"} + :CF_2 + SO_2 + CO_2$$
$$F^- + :CF_2 + CuI \rightarrow CF_3CuI$$
$$CF_3CuI + S \rightarrow CF_3SCu + I$$
$$CF_3SCu + ArI \rightarrow CF_3SAr + CuI$$
$$FSO_2CF_2CO_2Me + CuI \rightarrow FSO_2CF_2CO_2Cu + MeI$$

In the presence of aryl iodides, trifluoromethylthioaromatics are produced, although there is again a solvent effect. While NMP is generally effective, DMF is generally ineffective (which can be contrasted with its successful use in trifluoromethylations based on "$CuCF_3$" — see Chapter 5) and HMPA (also effective in stabilizing $CuCF_3$) is only effective with the more reactive aryl iodides.

Reactions based on electrophilic substitution (e.g., those based on CF_3SCl and $(CF_3SO_2)_2O$) and radical reactions are often exothermic and should be run with cooling. The reaction of CF_3Br with thiophenoxides are typically carried out under moderate pressure. Inhibition of reaction by nitrobenzene is consistent with a $S_{RN}1$ mechanism.

The fluoride-catalyzed cross-coupling of arene sulfonyl fluorides with (trifluoromethyl)trimethylsilane and (trifluoromethyl)trimethylstannane should be carried out under an inert atmosphere (e.g., argon). In some cases the reaction is sensitive to the solvent and THF or petroleum ether are generally the solvents of choice. The yield is dependent on the nature of the fluoride catalyst and for a given reaction can vary from 10 to 60%. The catalyst TASF, $(Me_2N)_3S^+Me_3SiF_2^-$, gives more consistently high yields than others, such as tetrabutylammonium fluoride (non-fluoride-soluble nucleophilic catalysts such as quaternary phosphonium phenoxides may also be used but can be much less active).

The conversion of aromatic disulfides to the corresponding trifluoromethylthioaromatics using trifluoromethyl halides requires the generation of the trifluoromethyl radical, CF_3. These can conveniently be produced using a sulfur dioxide radical anion precursor, such as $Na_2S_2O_4$, or a mixture of SO_2 and a reductant such as zinc metal:

$$CF_3Br + SO_2^- \rightarrow CF_3 + Br^- + SO_2$$
$$ArSSAr + CF_3 \rightarrow ArSCF_3 + ArS$$

This tends to be more efficient than photochemical activation. Remarkably, electron-rich aromatics do not require stoichiometric quantities of the reductant, 0.1 mol equivalent being sufficient. This can be explained by reduction of the formed SO_2 back to its radical anion by an intermediate cyclohexadienyl radical.[35,42,43]

REFERENCES

1. L.M. Yagupolsky and M.S. Maranets, *J. Gen. Chem., USSR,* 1955, 25, 1725.
2. E.A. Nocliff, S. Lipschutz, P.N. Craig and M. Gordon, *J. Org. Chem.,* 1960, 25, 60.
3. L.M. Yagupolsky, in *Aromatic and Heterocyclic Compounds with Fluorine-Containing Substituents,* Nasukova Dumka, Kiev, 1989, p.132.
4. *Organofluorine Chemistry: Principles and Commercial Applications,* eds. R.E. Banks, B.E. Smart and J.C. Tatlo, Plenum Press, New York, 1994.
5. B. Langlois in *Fluorine in Agriculture,* ed. R.E. Banks, Fluorine Technology, Cheshire, 1994.
6. J.L. Clave, B. Langlois, C. Wakselman and M. Tordeux, *Phosphorus, Sulphur Silicon,* 1991, 59, 129.
7. W.A. Sheppard, *J. Am. Chem. Soc.,* 1963, 83, 1314.
8. *Thermotropic Liquid Crystals,* ed. G.W. Gray, Wiley, New York, 1987.
9. M.A. McClinton and D.A. McClinton, *Tetrahedron,* 1992, 48, 6555.
10. A. Leo, C. Hansch and D. Elkins, *Chem. Rev.,* 1971, 71, 525.
11. W.A. Sheppard, *J. Org. Chem.,* 1964, 29, 1.
12. A.J. Beaumont and J.H. Clark, *J. Fluorine Chem.,* 1991, 52, 295.
13. J. Dickey, U.S. Patent, 2, 436, 100 (1938).
14. L.M. Yagupalsky and A.I. Kiprianov, *J. Gen. Chem. USSR,* 1952, 22, 2273.
15. I.G. Farbenindustrie A.-G, French Patent 820, 795 (1937); *Chem. Abstr.,* 1938, 32, 3422.
16. I.G. Farbenindustrie, A-G., British Patent 479, 774 (1938).
17. L.M. Yagupolsky and M.S. Marenets, *J. Gen. Chem. USSR,* 1954, 24, 885.
18. L.M. Yagupolsky, N.V. Kondratenko and V.P. Sambur, *Synthesis,* 1975, 721.
19. H.L. Yale, F. Sowinski and J. Bernstein, *J. Am. Chem. Soc.,* 1957, 79, 4375.
20. D.C. Remy, K.E. Rittle, C.A. Hunt and M.B. Freedman, *J. Org. Chem.,* 1976, 41, 1644.
21. J.H. Clark, C.W. Jones, A.P. Kbyett, M.A. McClinton, J.M. Miller, D. Bishop and R.J. Blade, *J. Fluorine Chem.,* 1990, 48, 249.
22. J.H. Clark and H. Smith, *J. Fluorine Chem.,* 1993, 61, 223.
23. N.V. Kondratenko, A.A. Kolomeytrev, V.I. Popov and L.M. Yagupolsky, *Synthesis,* 1985, 667.
24. Q.-Y. Chen and J-X. Duan, *J. Chem. Soc. Chem. Commun.,* 1993, 918.
25. A. Haas and W. Dmowski, *Chimia,* 1985, 39, 185; (Chem. Abstr., 1986, 104, 33982).
26. W. Dmowski and A. Haas, *J. Chem. Soc. Perkin Trans 1,* 1987, 2119.
27. W. Dmowski and A. Haas, *J. Chem. Soc. Perkin Trans 1,* 1988, 1179.
28. G.A.R. Brandt, H.J. Emeleus and R.N. Haszledine, *J. Chem. Soc.,* 1952, 2198.
29. A. Haas and C. Klare, *J. Fluorine Chem.,* 1989, 42, 265.
30. S. Andreades, J.F. Harris, Jr., and W.A. Sheppard, *J. Org. Chem.,* 1964, 29, 899.
31. W.A. Sheppard, *J. Org. Chem.,* 1964, 29, 895.
32. R.N. Haszeldine, R.B. Rigby and A.E. Tipping, *J. Chem. Soc. Perkin Trans. 1,* 1972, 2180.
33. A. Haas, *Angew. Chem. Int. Ed. Engl.,* 1981, 20, 647.
34. C. Wakselman and M. Tordeux, *J. Org. Chem.,* 1985, 50, 4047.
35. C. Wakselman, M. Tordeux, J.-L. Clavel and B. Langlois, *J. Chem. Soc. Chem. Commun.,* 1991, 993.
36. J.B. Hendrickson and K.W. Bair, *J. Org. Chem.,* 1977, 42, 3875.
37. L.M. Yagupolsky, N.V. Kondratenko and V.P. Sambur, *Synthesis,* 1975, 249.
38. C. Walsh, *Tetrahedron,* 1982, 38, 871.
39. A. Sekiya and T. Umemoto, *Chem. Lett.,* 1982, 1519.
40. S.J. Temple, *J. Org. Chem.,* 1968, 33, 344.
41. O.R. Pierce, E.T. McBee and G.F. Judd, *J. Am. Chem. Soc.,* 1954, 76, 477.
42. C. Wakselman and M. Tordeux, *J. Chem. Soc. Chem. Commun.,* 1987, 1701.
43. M. Tordeux, B. Langlois and C. Wakselman, *J. Chem. Soc. Perkin Trans. 1,* 1990, 2293.

Chapter 7

Other Aromatic Ring Substituents

7.1 INTRODUCTION

In previous chapters, synthetic strategies towards, and the properties of, aromatics containing the most important fluorinated substituents, namely fluorine itself and the trifluoromethyl group, have been discussed. The trifluoromethylthio and trifluoromethylsulfonyl groups have also been discussed. Other ring substituents, although less well studied, also have an important role in modern aromatic fluorine chemistry. Of these, the trifluoromethoxy, perfluoroalkyl, and fluoroalkyl side chains are probably the most important. There is currently a growing interest in the preparation of aromatics containing these groups for a number of applications, including the polymer, agrochemical, and pharmaceutical industries. These fluorinated compounds often show many of the enhanced properties that result from aromatic fluorination or trifluoromethylation. The major drawbacks associated with the incorporation of fluorinated ring substituents other than the trifluoromethyl group have been either the expense of the reagents involved, poor yields achieved, or the harsh reaction conditions required.

Trifluoromethoxy-substituted aromatics have found particular applications in the pharmaceutical and agrochemical industries. The major reason for this is undoubtedly the high lipophilicity of the trifluoromethoxy group, which is comparable to that of the trifluoromethyl and trifluoromethylthio groups. Routes to trifluoromethoxy-substituted aromatics are extremely limited, and in general the compounds are prepared by either the fluorination of the trichloromethoxy analogue or the reaction of a phenol with carbon tetrachloride, both of which require forcing conditions.

A number of strategies towards perfluoroalkyl-substituted aromatics are available, many of which are simply analogues of the trifluoromethylation reactions. Thus, perfluoroalkylation using perfluoroalkyl copper reagents, or the perfluoroalkylation of existing alkyl side chains, for example, are both possible. Perfluoroalkyl-substituted aromatics find applications in polymer and material chemistry, where the hydrophobic properties of the perfluoroalkyl chain can significantly enhance the overall hydrophobicity of the molecule, as well as the pharmaceutical and agrochemical industries, where the lipophilicity is again a dominant factor. Partially fluorinated side chains, such as $-OCF_2H$, also show some of the advantages as their perfluorinated analogues, and may be useful where some residual side-chain activity is required.

7.2 PHYSICAL PROPERTIES

The importance of the physical size of fluorinated substituents, as previously discussed, is of prime importance. The trifluoromethyl group (radius 2.7 Å) can be substituted for the methyl group (2.0 Å radius) with little disruption in steric geometry, and so the methoxy group can also be replaced by the trifluoromethoxy group with minimal disruption. The effects of the introduction of perfluoroalkyl chains are more significant, however. The volume of the perfluoroalkyl chain is significant, and it is likely that this can affect the steric geometry of the whole molecule, particularly for larger chains.

As with fluorine, trifluoromethyl, and trifluoromethylthio substituents, the fluorinated substituents have extremely high electronegativities, which have been previously listed, in Table 5.3. Replacement of a OCH_3 group by OCF_3 generally results in a decrease in boiling and melting points and an increase in density.

Of particular importance in biological applications is the lipophilicity of a molecule. High lipophilicity often results in a significant increase in the physiological effects of a molecule.[2] The trifluoromethoxy group, as with the trifluoromethyl and trifluoromethylthio groups, is highly lipophilic; the Hansch 1-octanol/water partition coefficients have been determined for a number of fluorinated groups and are given in Table 7.1.[3] In general, replacement of a fluorine by bromine reduces the lipophilicity, and replacement by hydrogen reduces this further. The incorporation of oxygen, and particularly sulfur, between the ring and the fluoroalkyl group increases the lipophilicity significantly.

Table 7.1 Hansch Hydrophobic Parameters for a Range of Fluorinated Groups

Group	Hydrophobic Parameter
SCF_3	1.44
SCF_2Br	1.28
OCF_3	1.04
OCF_2Br	1.01
CF_3	0.88
SCF_2H	0.68
OCF_2H	0.58

From I. Roco and C. Wakselman, *Tetrahedron Lett,* 1981, 22, 323. With permission.

Perfluoroalkyl groups are also highly electronegative, and have similar electronic effects as trifluoromethyl groups. Since these groups often find applications in polymer chemistry, the hydrophobic properties are also of interest. Perfluoroalkyl groups act as highly hydrophobic chains pendant to a molecule and often affect the hydrophobic properties of the entire molecule.

7.3 CHEMICAL PROPERTIES

The trifluoromethoxy group, like the trifluoromethyl group, is relatively inert, and is unaffected by mild acids, bases, and organometallic reagents.[4] The trifluoromethoxy group has a similar inductive and resonance effect as the halogens, and has been described as a "super halogen".[1] The group deactivates aromatic rings towards further electrophilic substitution, and is strongly para-orienting, having been studied in a range of reactions, including nitrations, acylations, carbonylations, and sulfonylations.[5,6] Unlike the strongly electron-withdrawing trifluoromethyl group, the trifluoromethoxy group does not activate aromatics towards aromatic nucleophilic substitution reactions without copper (I) catalysis. Meta isomers of trifluoromethoxy-substituted aromatics can only be prepared by nucleophilic substitution via the reaction of a *m*-trifluoromethoxy halo aromatic with sodium and liquid ammonia via a benzyne intermediate.[7]

7.4 SYNTHESIS OF TRIFLUOROMETHOXY-SUBSTITUTED AROMATICS

Although the trifluoromethoxy group has a similar effect on the properties of a molecule as the trifluoromethyl and trifluoromethylthio groups, few methods exist for its incorporation. In general, these methods require harsh reaction conditions, which in many cases limit the range of suitable substrates. Nevertheless, because of the properties of the trifluoromethoxy group, a number of important biologically active compounds containing this group have been prepared. Riluzole (2-amino-6-trifluoromethoxybenzotriozole) has been widely studied as an anticonvulsant.[8] Trifluoromethoxy-substituted 4-quinolones (**I**), for example, have potential as cardiovascular drugs, requiring significantly lower doses to achieve a reduction in blood pressure as their nonfluorinated analogues.[9]

(I)

Whereas the trifluoromethyl and trifluoromethylthio groups may be directly incorporated into an aromatic ring, for example via the reaction of a trifluoromethyl copper reagent with a halogen-substituted aromatic, no analogous methods exist for the direct introduction of the trifluoromethoxy group. Although trifluoromethyl hypofluorite will allow the introduction of the trifluoromethoxy group into carbohydrates or alkenes, reaction with aromatic substrates gives only poor conversion to the trifluoromethoxy aromatic.[10] Trifluoromethoxy aromatics are therefore synthesized by modification of existing functional groups, primarily via two routes:

- Halogen exchange reactions of α,α,α-trichloromethoxy aromatics
- Conversion of phenols by reaction with carbonyl difluoride/sulfur tetrafluoride or carbon tetrachloride/hydrogen fluoride

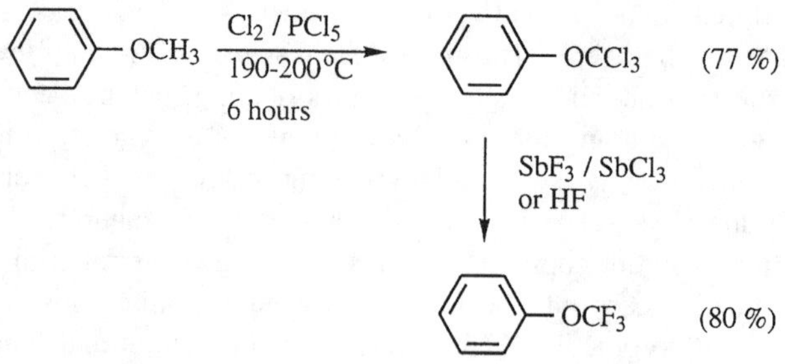

FIGURE 7.1   Chlorination and Swartz halogen exchange of trifluoromethoxy-substituted aromatics.

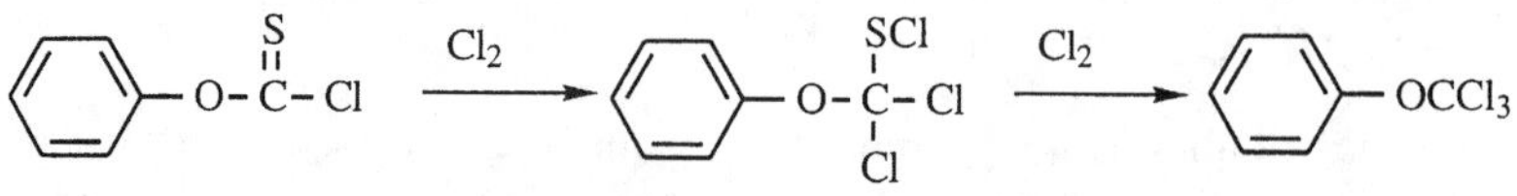

FIGURE 7.2   Chlorination of phenyl esters of chlorothiocarbonic acid.

7.4.1   Halogen Exchange Reactions of α,α,α-Trichloromethoxy Aromatics

The major drawback to this process is the availability of the α,α,α-trichloromethoxy-substituted aromatic precursor.[11] Trichloromethoxybenzenes are generally prepared from the corresponding anisole by reaction with chlorine or phosphorus pentachloride. The trichloro intermediate is then fluorinated via a Swartz halogen exchange procedure to give the trifluoromethoxy-substituted aromatic (Figure 7.1).[4] Since the reaction conditions are harsh, the viability of the reaction is limited to substrates bearing only halogen and nitro substituents.

An alternative synthesis of trichloromethoxy aromatics involves the reaction of the phenyl esters of chlorothiocarbonic acid with chlorine[12] (Figure 7.2).

7.4.2   Conversion of Phenols to Trifluoromethoxybenzenes

Since phenols are widely available, this route is particularly useful for the synthesis of trifluoromethoxy-substituted aromatics. Heating the phenol with carbon tetrachloride and hydrogen fluoride to 150°C in a sealed tube gives reasonable yields of the trifluoromethoxy aromatics[13] (Figure 7.3). It is possible that the reaction actually proceeds via a trichloromethoxy aromatic intermediate, which is generated *in situ* and subsequently fluorinated.

Phenols will react with carbonyl fluoride to give the fluoroformate derivatives, which can subsequently be converted to the trifluoromethoxy aromatic with sulfur tetrafluoride in the presence of catalytic amounts of anhydrous hydrogen fluoride[1] (Figure 7.4). This method also provides a convenient route for the synthesis of perfluoroalkoxy-substituted aromatics. Trifluoroacetate and heptafluorobutyrate esters of phenols also react with sulfur tetrafluoride under similar conditions,

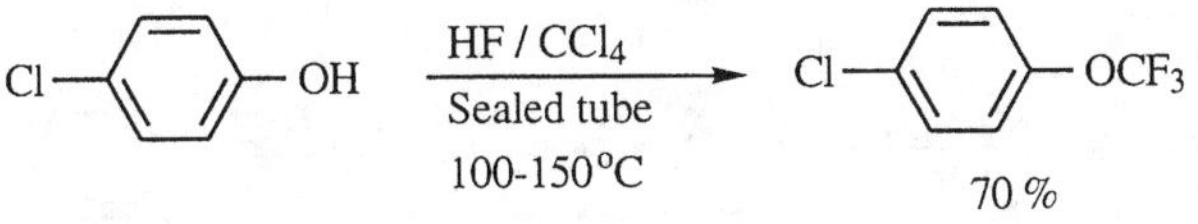

FIGURE 7.3 Reaction of phenols with carbon tetrachloride and HF.

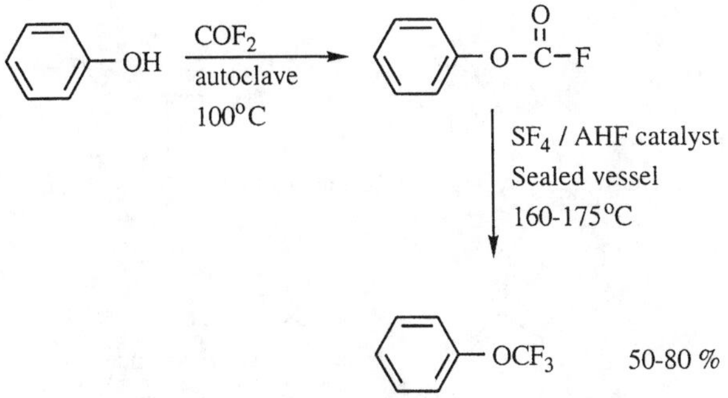

FIGURE 7.4 Conversion of phenols to trifluoromethoxybenzenes via reaction with carbonyl fluoride and sulfur tetrafluoride/HF.

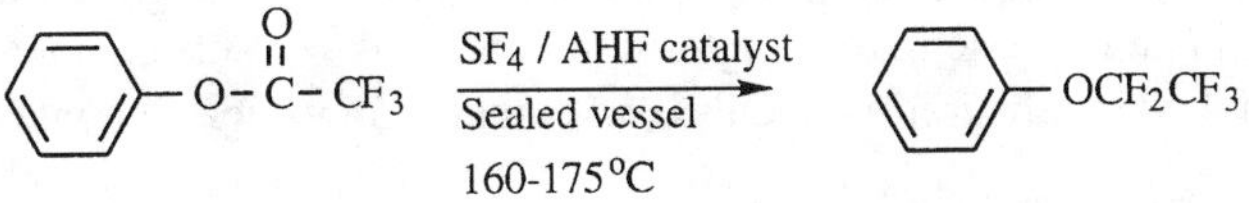

FIGURE 7.5 Synthesis of pentafluoroethoxy-substituted aromatics.

although yields of perfluoroalkoxy aromatics generally decrease as the length of the chain increases (Figure 7.5).

Many trifluoromethoxy-substituted aromatics that show biological activity are derived from trifluoromethoxy-substituted anilines. Since the group directs ortho and para for electrophilic substitution, the synthesis of m-trifluoromethoxyaniline is not straightforward. Langlois and Soula introduced a procedure for the synthesis of this potentially important intermediate, which they describe as being simple, safe, and realistic for the plant-scale production of trifluoromethoxy-substituted aromatics.[14] The process involves the methylation of o-chlorophenol using dimethyl sulfate to give the o-chloroanisole. The anisole is then chlorinated either with chlorine and ultraviolet irradiation, or with chlorine and a phosphorus pentachloride catalyst. Swartz halogen exchange is then carried out using the usual procedure. The final, key step is a regiospecific arynic amination reaction (Figure 7.6).

7.5 SYNTHESIS OF FLUOROALKOXY-SUBSTITUTED AROMATICS

Fluoroalkoxy-containing drugs, such as Flecainide Acetate (3M Pharmaceuticals), an antiarrythmic, are already marketed.[9]

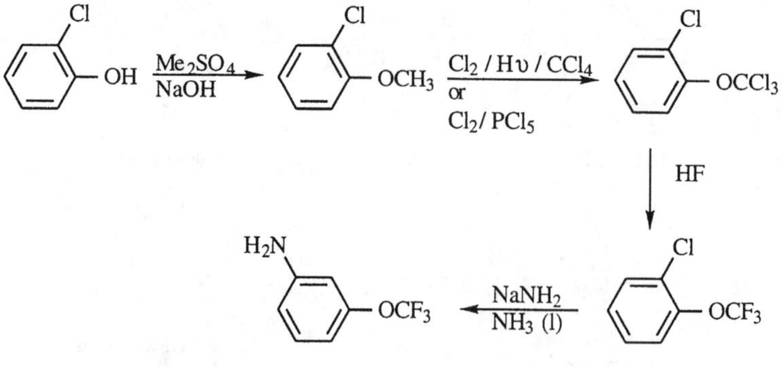

FIGURE 7.6   Synthesis of 3-amino-trifluoromethoxybenzene.

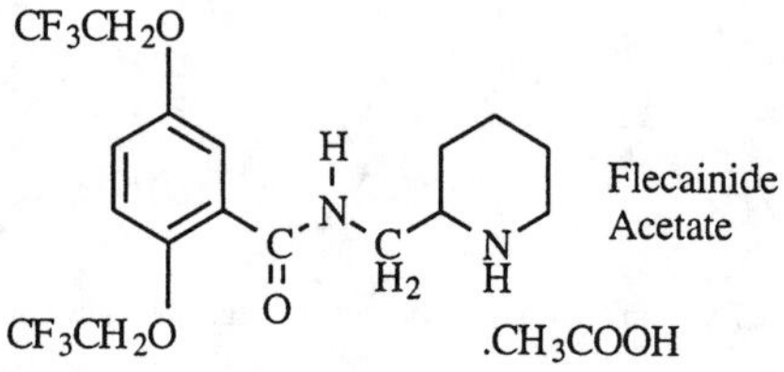

Flecainide
Acetate

.CH$_3$COOH

Fluoroalkoxy-substituted polyimides show a decrease in their glass transition temperatures with an increase in the length of the fluoroalkoxy side chain, and dielectric constants and water absorption decrease with increasing fluorine content in the side chain.[15]

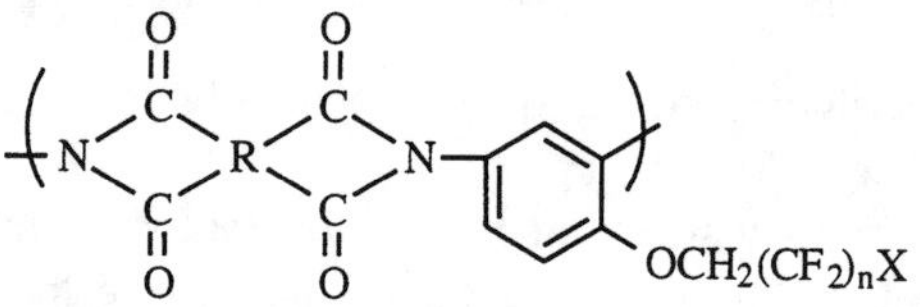

R = Connecting group
n = 3 - 10
X = F or H

The earliest methods available for the synthesis of such compounds involved the reaction of a phenol with an electrophilic haloalkyl fluoride or fluoroalkyl sulfonate, and required extended heating with either potassium carbonate or bicarbonate[16] (Figure 7.7).

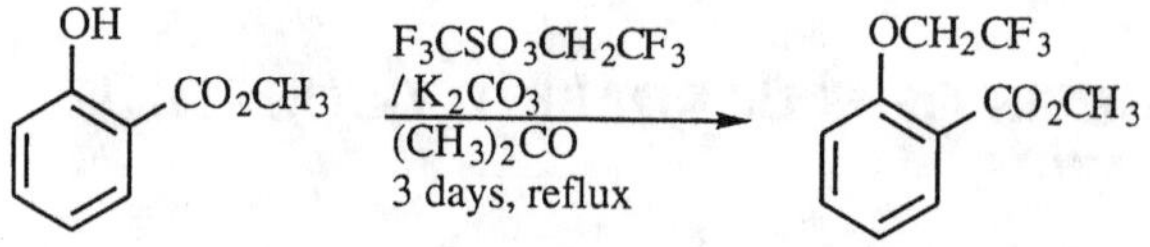

FIGURE 7.7   Fluoroalkylation of phenol.

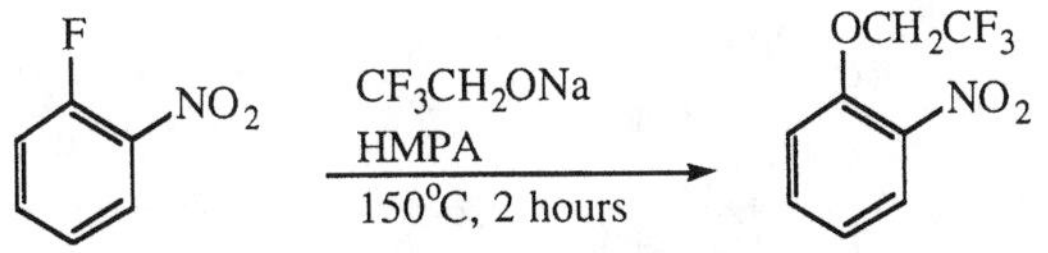

FIGURE 7.8 Nucleophilic substitution routes to fluoroalkyl-substituted aromatics.

More recently, direct nucleophilic aromatic substitution reactions have been developed, which permit fluoroalkoxy groups to be incorporated into a range of activated aromatics. The reaction has been extended to both halogen-substituted monoaromatic and heterocyclic substrates[17] and nitro-substituted aromatics[18] (Figure 7.8). In general, dipolar aprotic solvents such as dimethyl sulfoxide (DMSO) or HMPA are required to promote the reaction between the substrate and the sodium salt of an alkoxide.

7.6 SYNTHESIS OF PERFLUOROALKYL-SUBSTITUTED AROMATICS

Despite the growing interest in the use of aromatics containing perfluoroalkyl groups for a number of applications, including pharmaceuticals, agrochemical, and polymers, no perfluoroalkyl-substituted aromatics are commercially available in research quantities at present. Many of the routes to trifluoromethyl-substituted aromatics discussed in Chapter 5 are equally applicable for the synthesis of aromatics containing larger perfluoroalkyl groups. In addition, a number of other routes that have not been used for trifluoromethylation have been developed for perfluoroalkylation. The synthetic routes may be broadly divided into three categories:

- Direct electrophilic/radical incorporation of a perfluoroalkyl group
- Fluorination of an existing alkyl side chain
- Nucleophilic substitution reactions

7.6.1 Direct Incorporation of Perfluoroalkyl Groups

Heating aromatic substrates with perfluoroalkyl iodides in a sealed reaction vessel generates perfluoroalkyl-substituted aromatics via a free radical process[19] (Figure 7.9). The reaction of polyfluoroacyl chlorides with aromatic substrates in the presence of nickel tetracarbonyl also gives perfluoroalkyl-substituted aromatics[20] (Figure 7.10). Both of these processes require the use of two stoichiometric equivalents of the perfluoralkyl reagent, which is generally expensive. The harsh reaction conditions required are also unsuitable for a number of ring substituents.

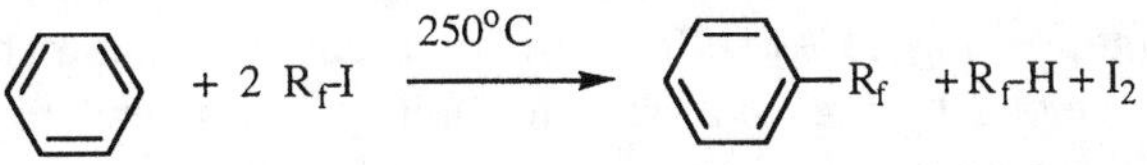

FIGURE 7.9 Reaction of perfluoroalkyliodides with benzene.

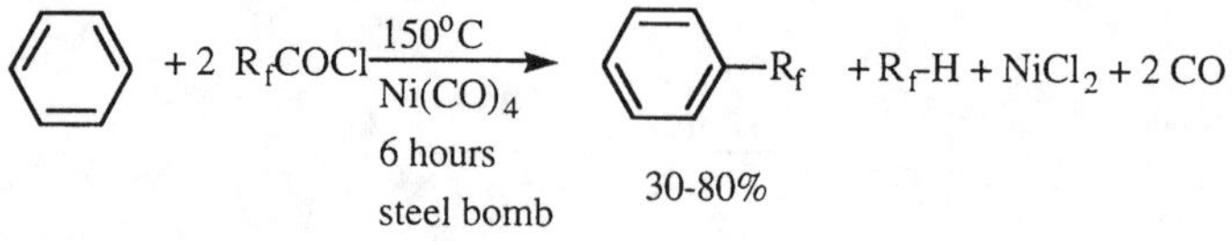

FIGURE 7.10   Reaction of polyfluoroacyl chlorides with benzene.

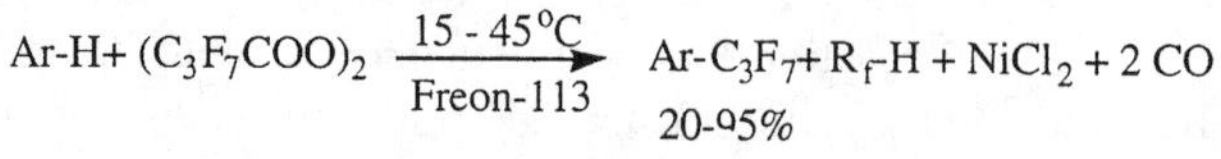

FIGURE 7.11   Reaction of bis(perfluoroalkanoyl) peroxides with benzene.

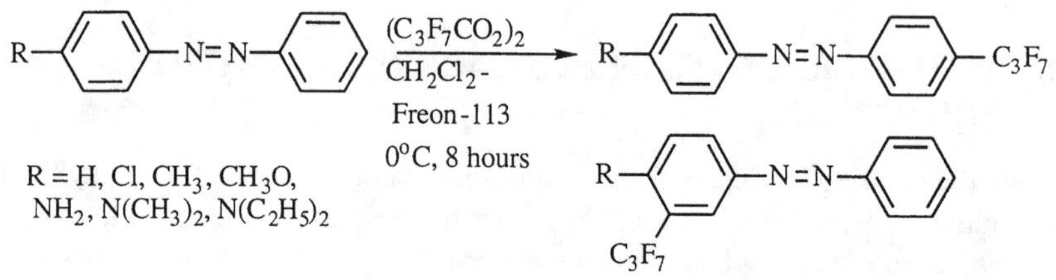

R = H, Cl, CH$_3$, CH$_3$O,
NH$_2$, N(CH$_3$)$_2$, N(C$_2$H$_5$)$_2$

3-40 % yield of perfluoroalkylated
products

FIGURE 7.12   Heptafluoropropyl-substituted azobenzene dyes.

Bis(perfluoroalkanoyl) peroxides will react under relatively mild conditions with electron-rich benzenes,[21] including nitrogen-containing aromatics (pyrroles but not pyridines)[12] and substituted thiophenes.[23] The peroxide, bis(heptafluoropropanoyl)peroxide, is prepared from heptafluorobutyryl chloride and hydrogen peroxide in 1,1,2-trichloro-1,2,2-trifluoroethane (Freon-113) and is used directly for the perfluoroalkylation reaction (Figure 7.11). This system has been used for the synthesis of heptafluoropropyl-substituted azobenzenes.[24] The resulting dyes show bathochromic shifts of 20 to 90 nm, more than their nonperfluoroalkylated analogues (Figure 7.12).

Arylperfluoroalkyliodinium chlorides, prepared by the reaction of bis(trifluoroacetoxy)iodoperfluoroalkanes with toluene in trifluoroacetic acid, will react with monomethyl and dimethylanilines to give *para*-perfluoralkyl-substituted aromatics under mild conditions, rather than the expected nucleophilic attack by the nitrogen, which occurs with nonmethylated anilines[25] (Figure 7.13).

Sodium dithionite has been reported as an initiator for the radical reaction of perfluoroalkyl halides (chlorides, bromides, and iodides) with aromatic and heteroaromatic substrates[26] (Figure 7.14). The use of a cationic phase transfer catalyst extends the range of substrates to unactivated aromatics, although di-perfluoroalkyl-substituted biphenyls may also be produced in competition.

The direct perfluoroalkylation of both aromatic and heteroaromatic substrates with trifluoromethyl or perfluorohexyl sulfonyl chlorides has recently been reported.

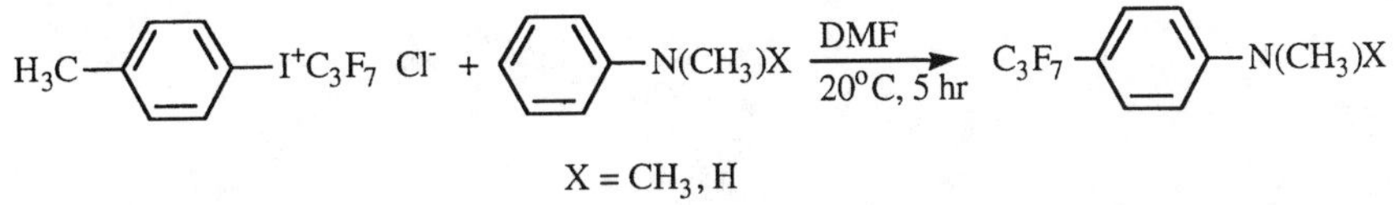

FIGURE 7.13 Aromatic perfluoroalkylation using arylperfluoroalkyliodinium chlorides.

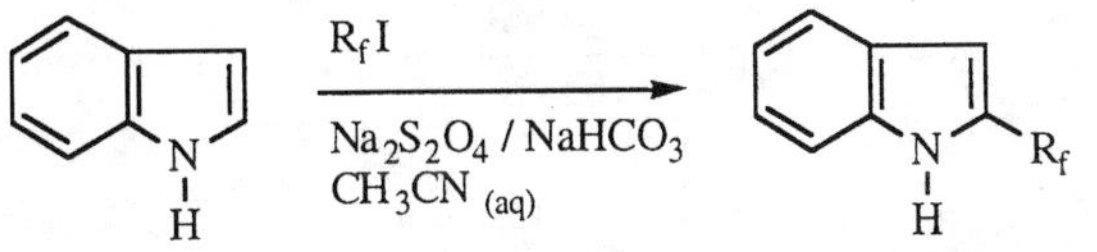

FIGURE 7.14 Sodium dithionite-initiated aromatic perfluoroalkylation.

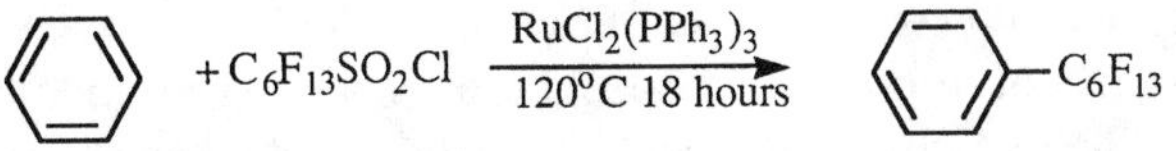

FIGURE 7.15 Aromatic perfluoroalkylation catalyzed by ruthenium (II) phosphine complexes.

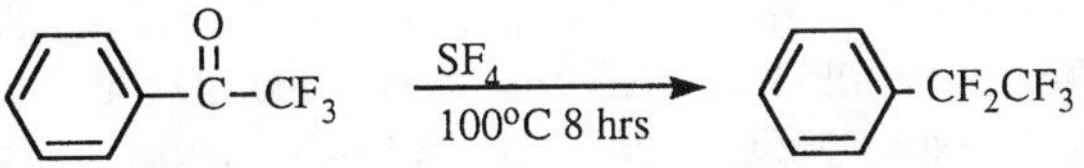

FIGURE 7.16 Fluorination of fluoroalkyl aryl ketones with sulfur tetrafluoride.

In this system, a ruthenium (II) phosphine complex is used as a catalyst. The procedure is not suitable for substrates possessing strong electron-withdrawing groups, such as nitrobenzene[27] (Figure 7.15).

7.6.2 Fluorination of Existing Side Chains

The attempted fluorination of alkyl benzenes generally gives either competitive ring fluorination or ring oxidation. Fluoroalkyl arylketones may, however, be successfully fluorinated using sulfur tetrafluoride[28] (Figure 7.16). In general, the harsh reaction conditions required have limited this route for the production of perfluoroalkyl-substituted aromatics, since other functional groups present are also liable to undergo reaction.

7.6.3 Nucleophilic Substitution Reactions

Necleophilic substitution reactions account for the majority of examples of aromatic perfluoroalkylation reported in the primary literature in recent years. The design of pharmaceuticals, for example, often requires that the perfluoroalkylation stage is towards the end of a multistep synthesis, so that the other steps are not disrupted by the strong electronic effects of the perfluoroalkyl group. It is vital, therefore, that

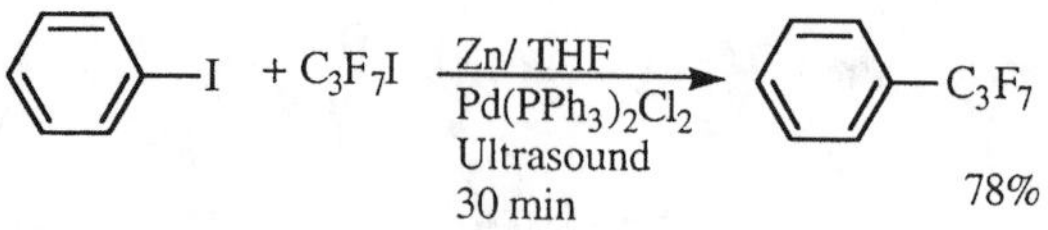

FIGURE 7.17   Pd (II)-catalyzed aromatic perfluoroalkylation.

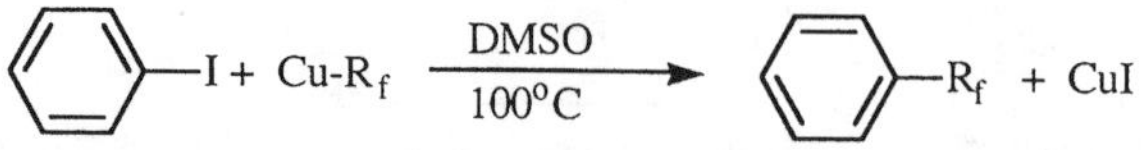

FIGURE 7.18   Aromatic perfluoroalkylation using perfluoroalkyl copper complexes.

perfluoroalkylation systems are developed which give both high yields and high selectivity under conditions that will not affect other functional groups in the substrate. Nucleophilic substitution reactions offer this possibility, and in general, bromo or iodo aromatics are used for the substitution.

Palladium has been used to catalyze a cross-coupling reaction between a perfluoroalkyl iodide and allyl, vinyl, and aryl halides.[29] Among the palladium complexes studied, $PdCl_2$ and $Pd(PPh_3)_2Cl_2$ were found to be most suitable. The reaction requires the use of zinc to generate a R_f-Zn-I type complex in THF promoted by ultrasound. The reaction proceeds at room temperature with high conversions being reported (Figure 7.17).

The use of perfluoroalkyl copper complexes is the most commonly cited route for aromatic perfluoroalkylation in recent years. Perfluoroalkyl copper reagents were first introduced by McLoughlin and Thrower in 1969, and used for the perfluoroalkylation of a wide range of iodoaromatics.[30] The copper complex is prepared by heating a perfluoroalkyl iodide with two equivalents of copper metal in DMSO for several hours. Reaction with an aryl iodide then generates the perfluoroalkyl-substituted aromatic (Figure 7.18).

The reaction conditions allow a wide variety of functional groups, including halide, nitro, nitrile, alkoxy, hydroxy, and carboxylate ester to be present in the substrate. Although DMSO is the preferred solvent, a number of other dipolar aprotic solvents, including sulfolane, DMF, and DMAc, are also suitable. The preparation of the perfluoroalkyl copper complex is generally carried out *in situ,* although in reactions involving highly activated substrates, competitive copper promoted reactions, including Ullmann couplings (to give biphenyls) and reductive dehalogenations may occur in competition. In such cases, the pregeneration of the perfluoroalkyl copper complex may then be advantageous. Unlike the trifluoromethyl copper complexes discussed in Chapter 5, longer-chain copper complexes can be isolated. Perfluoroheptyl copper can be isolated as a clear green syrup which decomposes slowly, although solutions in DMSO are stable indefinately, and those in ether decompose only slowly (about 1% per day).

Reactions of perfluoromethyl, ethyl, propyl, and butyl iodides are generally carried out in sealed tubes, since their boiling points are much lower than the reaction temperatures used (80 to 150°C). Perfluoroalkyl bromides rather than iodides are also used, although longer reaction times or higher reaction temperatures are generally required. Iodinated aromatics are often expensive or difficult to manufacture, and in recent years the range of substrates have been extended to include bromoaromatics,[31] bromoheterocyclics,[32] and halogenated polycyclics.[33]

   Labadie and colleagues have used the perfluoroalkyliodide/copper metal system to synthesize novel perfluorinated polyethers containing perfluoroalkyl chains in either the polymer backbone[34] or pendant to the ring.[35]

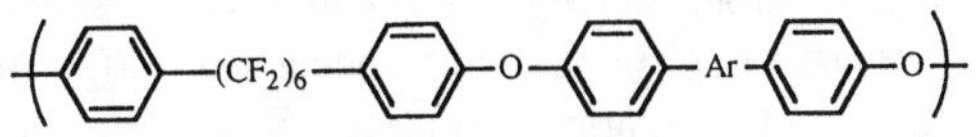

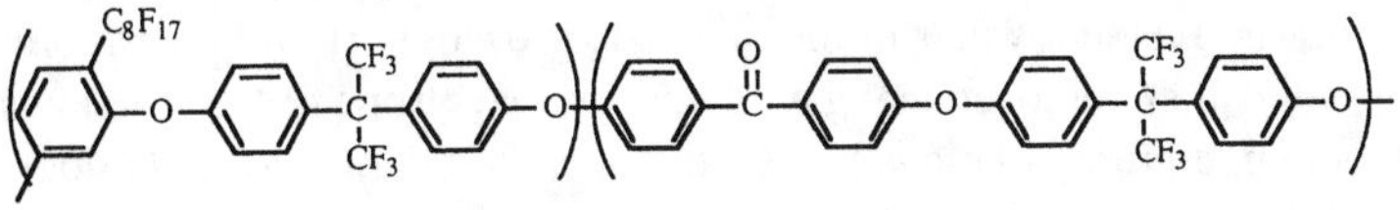

Ar = C(CH$_3$)$_2$ or C(CF$_3$)$_2$

This method has also been used to prepare *N*-[6-methoxy-5-(perfluoroalkyl)-1-naptholyl]-*N*-methylglycines, and their thionapthoyl analogues, which are active as reductase inhibitors and can be used to treat or prevent diabetic complications.[36] The heptafluoropropyl-substituted compounds show up to twice the physiological activity of the trifluoromethylated analogue.

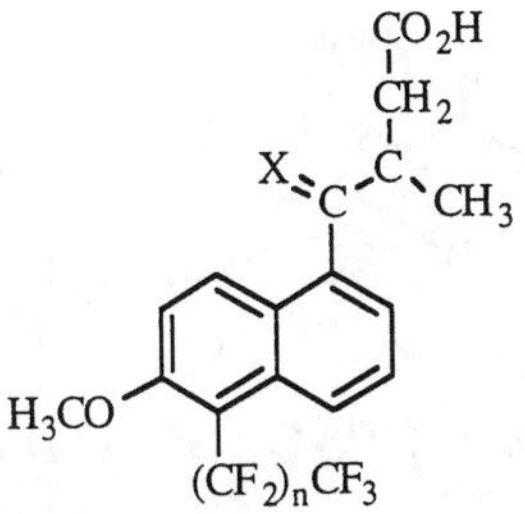

n = 1 to 5

X = O or S

   Other routes, other than the use of perfluoroalkyl iodides and copper metal, have also been discussed in Chapter 5. The Burton system, in which a CF_2X_2 (X = Br, Cl, I) compound is converted to a $CuCF_3$ species proceeds via a CF_2-type carbene intermediate. Pentafluoroethyl- and heptafluoropropyl-substituted products were seen in some cases, resulting from the reaction of $CuCF_3$ with CF_2. Burton and coworkers showed that it is possible to maximize yields of perfluoroalkyl copper complexes, which can be used to give reasonable yields of pentafluoroethyl-substituted aromatics. The metathesis reaction also occurs with zinc and cadmium, in addition to copper. Allowing a mixture of Cd or Zn CF_3 and copper (I) to warm to room temperature results in the generation of $CuCF_3$, which then slowly converts to CuC_2F_5, which can be used for aromatic pentafluoroethylation.[37]

   The most inexpensive reagents for aromatic perfluoroalkylation are the salts of perfluoroalkylcarboxylic acids. The use of sodium trifluoroacetate for aromatic

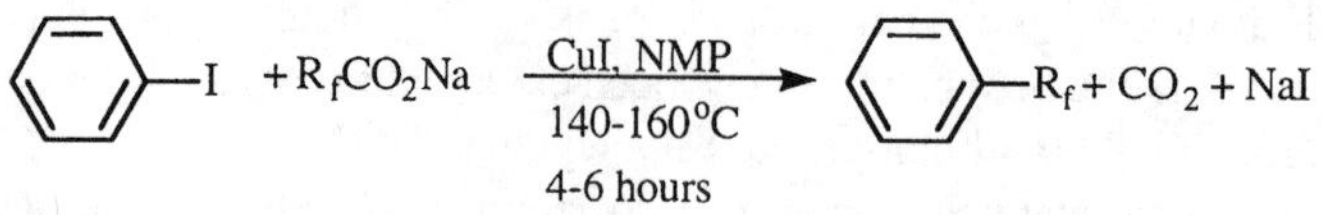

FIGURE 7.19   Aromatic perfluoroalkylation using perfluoroalkylcarboxylate salts.

trifluoromethylation has been discussed in Chapter 5, and this method has been shown to be equally suitable for larger perfluoroalkylcarboxylate salts. The reaction must be carried out at 140 to 160°C to promote the decarboxylation of the salt, but perfluoroalkyl complexes decompose at approximately 150°C. Therefore, up to four equivalents of the salt are used to allow for some decomposition during the reaction[38] (Figure 7.19).

In some cases, significant amounts of reductively dehalogenated or biaryl products result. The residual water in the highly hygroscopic sodium perfluoroalkylcarboxylate salt can be removed using a toluene pre-azeotrope of the reaction system.[39] Since this temperature is below 140°C, the decarboxylation reaction does not take place during this stage. Using this method, the amount of sodium salt and reaction time can be reduced considerably. BRL 55834 is an airway-selective potassium channel activator, which has applications as an antiasthmatic agent, and has been prepared by the reaction of a bromopyran with sodium pentafluoropropionoate using this procedure.[40]

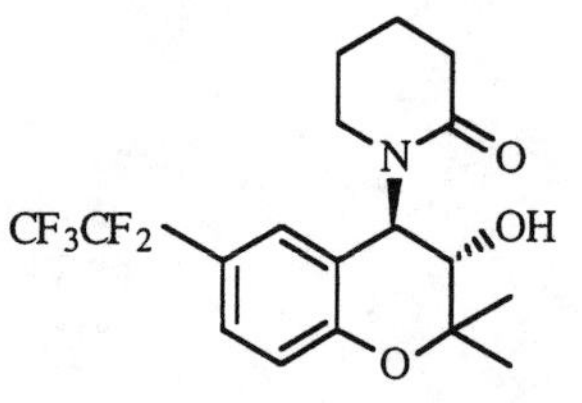

BRL 55834

7.7 SYNTHESIS OF FLUOROALKYL-SUBSTITUTED AROMATICS

Fluoroalkyl-substituted aromatics are particularly useful as intermediates, since the residual protons are strongly activated by the fluorine in the side chain. Toluenes containing electronegative substituents (such as nitro and cyano) may be readily fluorinated at the methyl group using liquid hydrogen fluoride in conjunction with one of a number of metal salts. Examples of metal salts suitable for the reaction include PbO_2, NiO_2, CoF_3, $Co(CH_3CO_2)_2$, and AgF_2. In general, a mixture of mono- and difluoromethyl-substituted products result[41] (Figure 7.20).

A much more convenient route to fluoromethyl-substituted aromatics is via the halogen exchange reaction of a benzyl bromide (Figure 7.21). The metal fluorides discussed for the aromatic halex reactions in Chapter 2 are equally suitable for substitution at the benzylic position. Thus, potassium fluoride with a crown ether,[42] spray-dried potassium fluoride,[43] alkali metal fluorides,[44,45] silver fluoride supported on calcium fluoride,[46] lead (II) fluoride, and a sodium halide[47] have all been used

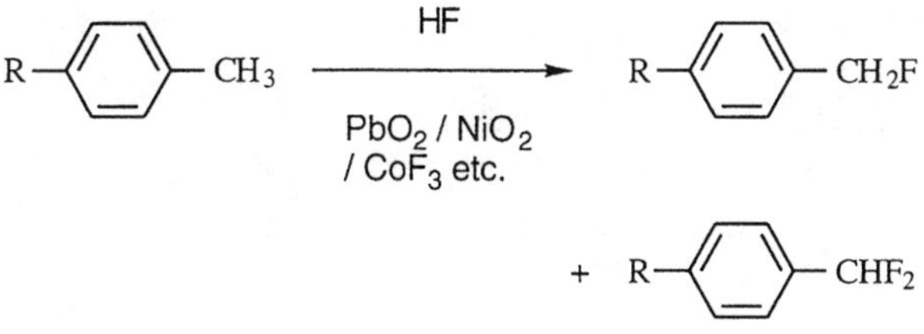

FIGURE 7.20 Side-chain fluorination of activated toluenes; R = electronegative group.

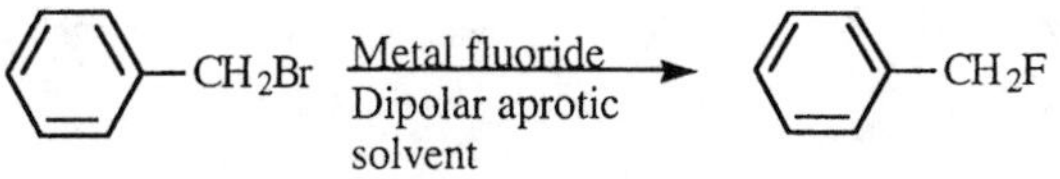

FIGURE 7.21 Halogen exchange fluorination of benzyl bromides.

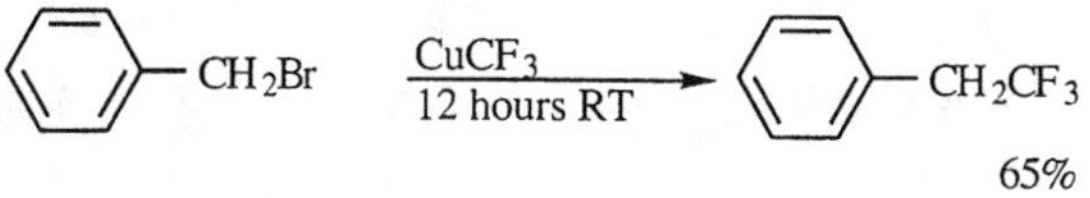

FIGURE 7.22 Reaction of benzylbromides with trifluoromethyl copper.

with success. A number of transition metal fluorides have been used for the reaction, and in general the order of reactivity is Zn (II) < Cu (II) > Co (II) > Mn (II). In this system, the reaction occurs readily in refluxing acetonitrile, although 2,2′-bipyridine is required as a complexing agent for the transition metal halide.[48]

The most common route to perfluoroalkyl-substituted aromatics (the reaction of a perfluoroalkyl copper complex with an iodoaromatic) is not suitable for 2,2,2-trifluoroethyl iodide giving poor conversion to the trifluoroethylaromatic.[49] Benzylbromides do, however, react with trifluoromethyl copper, providing a route to trifluoroethyl-substituted aromatics under relatively mild conditions[50] (Figure 7.22).

Other routes involve the Swartz halogen exchange reaction, and although competitive elimination reaction to give alkenes is often observed, up to 50% conversion to the desired product is possible[51] (Figure 7.23).

α,α,α-Trifluoroacetophenones have been reduced to the trifluororoethyl-substituted aromatics, although the procedure is complex; the acetophenone is prepared from a Grignard reagent, and reduced with sodium borohydride to the alcohol, which is then reduced to the alkane by reduction of the tosyl derivative of the alcohol[52] (Figure 7.24).

7.8 SYNTHESIS OF ARYL DIFLUOROMETHYL ETHERS AND THIOETHERS

Aryl difluoromethyl ethers and thioethers have been studied in some detail in recent years, particularly for pharmaceutical and agrochemical applications.[53] Although the stability of these groups is less than that of their perfluoro analogues, this may be

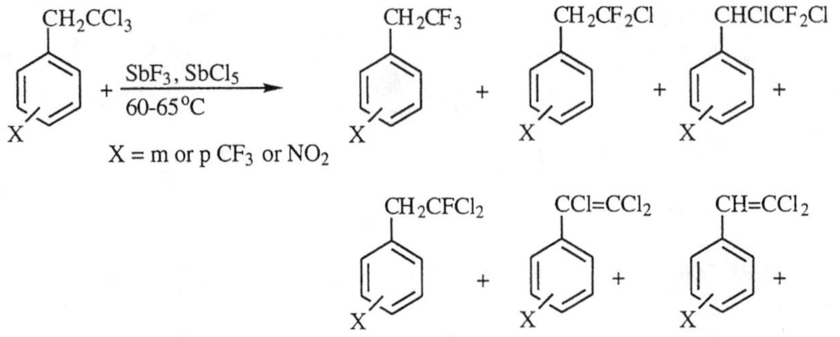

FIGURE 7.23 Swartz halogen exchange routes to fluoroalkyl-substituted aromatics.

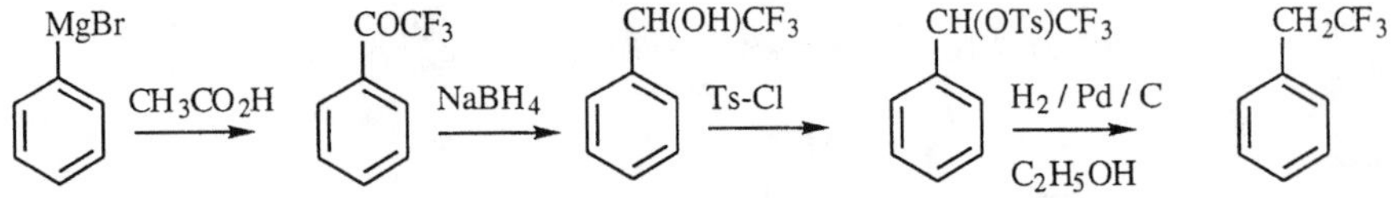

FIGURE 7.24 Grignard routes fluoroalkyl-substituted aromatics.

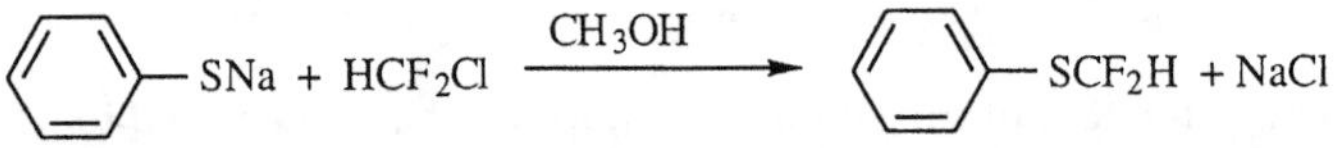

FIGURE 7.25 Reaction of alkali metal phenates with chlorodifluoromethane.

$$X = O, S$$

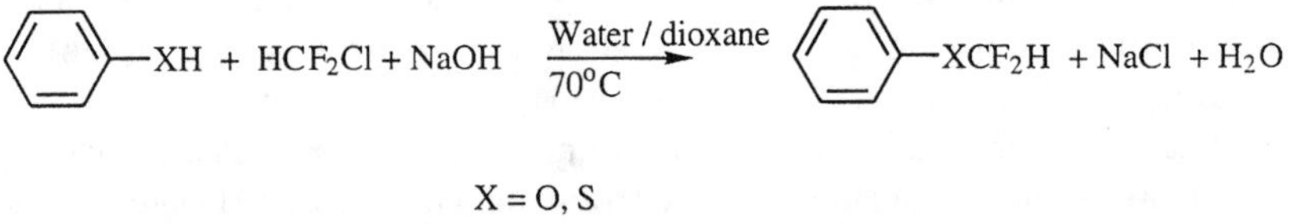

FIGURE 7.26 Reaction of phenol and thiophenol with chlorodifluoromethane and sodium hydroxide.

desirable to provide a site for further reactions, or where short-lived metabolites are required.

The first synthesis of these compounds was via the reaction of alkali metal phenates or thiophenates with chlorodifluoromethane. The reaction actually proceeds via a difluorocarbene intermediate[54,55] (Figure 7.25). The reaction has also been carried out using phenol or thiophenol with sodium hydroxide[56] (Figure 7.26).

Some drawbacks do exist with this procedure, notably hydrolysis of the difluorocarbene with high concentrations of base, and hydrolysis of the product during work up. The method has been modified through the use of a solid-liquid phase transfer.[53] The sodium hydroxide is used in a crushed form, limiting the concentration in solution. *Tris*-(3,6-dioxyheptyl)amine is used to transport the organic salts into the apolar aprotic solvent, such as carbon tetrachloride or an aromatic.

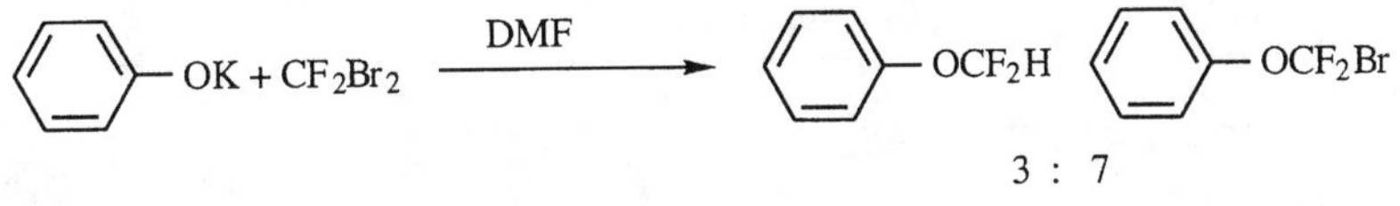

FIGURE 7.27 Reaction of phenates with dibromodifluoromethane.

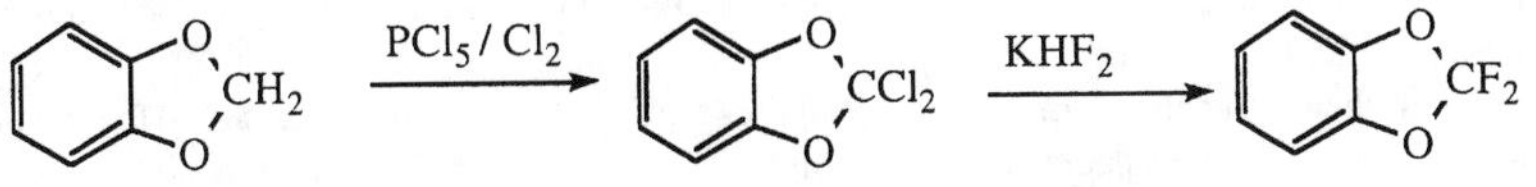

FIGURE 7.28 Fluorination of benzodioxazoles.

Bromodifluoromethyl and difluoromethyl ethers and thioethers have been prepared from the phenates and thiophenates via the reaction with CF_2Br_2. Again, the reaction proceeds via the difluoromethyl carbene; and bromodifluoromethyl- and difluoromethyl-substituted products are generally obtained[2] (Figure 7.27).

Benzodioxazoles are important ring systems occurring in natural products. Recently, difluoro derivatives have been prepared via the chlorination and halogen exchange reaction of the benzodioxazoles (Figure 7.28) to give a number of useful intermediates, including those for a commercial fungicide, Fludioxonie.[56] A number of fluorinating agents, including potassium bifluoride, antimony trifluoride, and hydrogen fluoride, have been used in this reaction system.

7.9 SYNTHETIC METHODS

7.9.1 Trifluoromethoxy-Substituted Aromatics

The major routes for the synthesis of trifluoromethoxy-substituted aromatics are summarized in Table 7.2. All of the methods in general require harsh conditions. The first route, from the methoxy aromatic is further hindered by the poor availability of starting materials. The chlorination of the anisole requires the use of chlorine itself or phosphorus pentachloride in an autoclave. The conversion of the phenyl esters of chlorothiocarbonic acid to the trichloromethoxy aromatic again is limited by the availability of suitable substrates. In each case, the fluorination step requires antimony trifluoride (highly toxic) and hydrogen fluoride, which presents particular handling difficulties. Again, the reaction requires the use of an autoclave, and is not particularly amenable to small-scale laboratory synthesis. The reaction is generally limited to nitro- and halo-substituted aromatics.

The other major alternative route, the conversion of phenols, is more attractive, due to the wider availability of the phenol substrates. The simplest route is via the one-step reaction with carbon tetrachloride and hydrogen fluoride. Alternatively, conversion to the anisole via reaction with dimethylsulfate (highly toxic and possibly carcinogenic) is then followed by chlorination and halogen exchange, giving a three-stage reaction. The other alternative, via the chloroformate derivative, requires the use of the highly toxic carbonyl fluoride, followed by sulfur tetrafluoride and anhydrous

Table 7.2 Synthetic Routes to Trifluoromethoxyaromatics

Route	Comments
Chlorination of anisoles and halogen exchange	Requires anisole precursor; 2-step reaction in autoclave with Cl_2/PCl_5 followed by SbF_3/HF; highly toxic/corrosive reagents
Direct conversion of phenols	Phenols readily available; 1-stage reaction with carbon tetrachloride and HF in sealed tube
Conversion of phenol to anisole followed by chlorination and halogen exchange	Requires dimethyl sulfate to give anisole (highly toxic, potential carcinogen); multistep route
Conversion of phenol to chloroformate derivative followed by fluorination with SF_4/AHF	Requires highly toxic carbonyl fluoride and the use of AHF (highly corrosive) and SF_4 (expensive, highly toxic gas); method can be readily extended to other perfluoroalkoxy-substituted aromatics

hydrogen fluoride. Sulfur tetrafluoride is expensive, highly toxic, gaseous (and therefore presents handling difficulties), and is readily hydrolyzed to HF.

7.9.2 Fluoroalkoxy-Substituted Aromatics

Original methods of synthesis, such as the reaction of a phenol with an electrophilic haloalkyl fluoride or fluoroalkyl sulfonate require forcing conditions, e.g., 3 days at reflux, and are also limited by the availability and cost of the reagents. Much more convenient routes are via direct nucleophilic substitution. Synthetic routes involving aromatic nucleophilic substitution and the use of dipolar aprotic solvents have been widely studied; for example, the halogen exchange chemistry discussed in a previous chapter.

7.9.3 Perfluoroalkyl-Substituted Aromatics

Again, many of the original methods, such as the direct reaction of perfluoroalkyl-idodides with aromatics are little used today. Newer methods, such as the palladium-catalyzed cross coupling of aryl halides with perfluoroalkyliodides, or the reaction of benzenes with bis(perfluoroalkanoyl) peroxides, also receive little attention. The most widely studied route is undoubtably the copper-mediated reaction of perfluoroalkyl iodides with iodoaromatics. Synthetic strategies using this route are summarized in Table 7.3. Perfluoroalkyl iodides are readily available, although highly expensive. Reactions using the lower homologues in the series, such as C_2F_5I, and C_3F_7I, are further complicated by their low boiling points (12 and 41°C, respectively). Copper and copper salts are readily available and inexpensive, unlike, for example, the palladium alternatives. Iodo aromatic precursors are also expensive, although readily available via diazotization of the aniline precursor. Bromo alternatives to both iodides may also be used, although reactions then require more forcing conditions. The simplest synthetic strategy is simply to mix the copper metal, perfluoro-alkyl idodide, and aryl iodide and heat. A dipolar aprotic solvent, typically DMSO, is required to stabilize organometallic intermediates. Since the intermediate perfluoroalkyl copper complexes are air and moisture sensitive, a dry inert atmosphere is

Table 7.3 Synthetic Routes to Perfluoroalkyl-Substituted Aromatics

Route	Comments
One-pot reaction of aryl iodide with Cu and R_f-I	Simplest route — suitable for low to medium activated substrates; requires expensive R_f-I
Decomposition of R_fCOONa *in situ*	Carboxylate salts inexpensive, easier to handle than R_f-I, but need up to 4 equivalents to counter decomposition; unsuitable for highly activated aromatics
Preformation of perfluoroalkyl copper complex	Useful for highly activated aromatics; may get decomposition before Ar-I added
Metathesis of difluorodihalomethane	Inexpensive (but volatile) CF_2X_2; only for trifluoromethyl and pentafluoroethyl aromatics

required. Sealed tubes are generally required for the low boiling perfluoroalkyliodides, together with efficient stirring and mixing. Pretreatment of the copper metal, by, for example, the action of dilute mineral acid, removes surface oxides and decreases the time required for reaction to commence.

A much cheaper source of the perfluoroalkyl group is the sodium perfluoroalkylcarboxylate salt (readily available by neutralization of the acid), which can be decomposed at 140°C *in situ* with copper iodide. The higher reaction temperature often leads, however, to increased copper-promoted side reactions, such as reductive elimination and cross coupling. Since the salt is a solid, the handling difficulties of the perfluoroalkyliodides are avoided. Side reactions resulting from traces of water may be reduced by pre-azeotroping the system, e.g., with toulene.

For highly active substrates, particularly those bearing an ortho or para nitro group, side reactions may be unavoidable. In this case, the perfluoroalkyl copper complex can be prepared before substrate is added. Since perfluoroalkyl copper complexes decompose at 140 to 150°C, the perfluoroalkylcarboxylate salt route is not particularly suitable.

The reaction of inexpensive difluorodihalomethanes, such as CF_2Br_2, with copper is limited to the synthesis of trifluoromethyl- and pentafluoroethyl-substituted aromatics, and in practice, it is likely that both will be formed during reaction, unless special precautions are taken to convert all "$CuCF_3$" complexes to their pentafluoroethyl analogues. As with the perfluoroalkyl iodides, volatility is again a problem.

7.9.4 Fluoroalkyl-Substituted Aromatics

A number of routes to the fluoroalkyl routes have been discussed. Many, such as the Grignard route, require multistep methodologies and are not particularly suitable for the synthesis of complex aromatic compounds. The fluorination of toluenes at the side chain gives a mixture of mono and difluoro products, which may be acceptable at the start of a synthetic cycle, but unacceptable in the final stages. A number of catalysts, including PbO_2, NiO_2, CoF_3, $Co(CH_3CO_2)_2$, and AgF_2, enable the reaction to be carried out in liquid hydrogen fluoride.

A much more viable route is via the halogen exchange of benzyl halides, monohalogenated derivatives of which are readily available. Benzal halides (α,α- dihalogenated) derivatives are also available. The technique uses the same well-established methodology as the halogen exchange (halex) reactions discussed in a previous chapter. Typically, KF (possibly with a crown ether catalyst) is used in a dipolar aprotic solvent such as DMSO. Other alkali metal fluorides and transition metal fluorides may also be used, and a balance between reactivity and cost of the metal salt will often be the deciding factor, as with the halex system. This system has the advantage that the metal fluoride system is much less corrosive than the liquid hydrogen fluoride system, enabling reactions to be performed using standard equipment.

7.9.5 Aryl Difluoromethyl Ethers and Thioethers

The simplest routes involve the reaction of phenates or thiophenates with the readily available and inexpensive chlorodifluoromethane to give difluoromethane ethers, or with dibromodifluoromethane to give difluorobromoethers. The (thio)phenates may be generated *in situ* by addition of sodium hydroxide to the system. The use of phase transfer catalysis in some cases limits hydrolysis of the products. With the dibromodifluoromethane system, however, reduction of the product to the difluoromethane ether is normally a side reaction.

REFERENCES

1. W.A. Sheppard, *J. Org. Chem.,* 1964, 29, 1.
2. T. Fujita, J. Iwasa and C. Hansch, *J. Am. Chem. Soc.,* 1964, 86, 5157.
3. I. Roco and C. Wakselman, *Tetrahedron Lett.,* 1981, 22, 323.
4. L.M. Yagupolskii, *Dokl. Akad. Nauk. S.S.S.R.,* 1953, 105, 100.
5. B. Blank, U.S. Patent 3,021,368 (1962) CA 1962, 57, 4597i.
6. B. Langlois, Paper 7, Fluorine in Agriculture, Manchester (UK), 1995.
7. B. Langlois and G. Soule, *Bull. Soc. Chim. Fr.,* 1986, 925.
8. N. Mitrani, F. Flamand, A. Uzan, J.J. Legrand, C. Gueremy and G. Lefur, *Neuropharmacology,* 1985, 24, 1085.
9. R.V. Davies, Conference Proceedings, Fluorine in Medicine in the 21st Century; UMIST, UK, 1994.
10. J.A.C. Allison and G.H. Cady, *J. Am. Chem. Soc.,* 1959, 81, 1089.
11. R. Louw and P.W. Franken, *Chem. Ind. (London),* 1977, 127.
12. British Patent 765,527, 1975.
13. A.E. Feiring, *J. Org. Chem.,* 1979, 44, 2907.
14. B. Langlois and G. Soula, *Bull. Chem. Soc. Fr.,* 1986, 925; *J. Fluorine Chem.,* 1987, 35, 39.
15. T. Ichino, S. Sasaki, T. Matsuura and S. Nishi, *J. Polym. Sci. Part A,* 1990, 28, 323.
16. A. Mendel, U.S. Patent 3766247: CA 1974, 80, 14747n.
17. J.P. Idoux, J.T. Gupton, C.K. McCurry, A.D. Crews, C.D. Jurss, C. Colon and R.C. Rampi, *J. Org. Chem.,* 1983, 48, 3771.
18. J.P. Idoux, M.L. Madenwald, B.S. Garcia and D.L. Chu, *J. Org. Chem.,* 1985, 50, 1876.
19. G.V. Tiers, *J. Am. Chem. Soc.,* 1960, 82, 1960.
20. J.J. Drysdale and D.D. Coffman, *J. Am. Chem. Soc.,* 1960, 82, 5111.
21. M. Yoshida, H. Amemiya, M. Kobayashi, H. Sawada, H. Hagii and K. Aoshima, *J. Chem. Soc. Chem. Commun.,* 1985, 234.
22. M. Yoshida, T. Yoshida, M. Kobayashi and N. Kamigato, *J. Chem. Soc. Perkin Trans. 1,* 1989, 909.
23. M. Yoshida, T. Yoshida, N. Kamigato and M. Kobayashi, *Bull. Chem. Soc. Jpn.,* 1988, 61, 3549.
24. M. Matsui, S. Kawamura, K. Shibita, M. Mitani, H. Sawada and M. Nakayama, *J. Fluorine Chem.,* 1992, 57, 209.

25. L.M. Yagupolskii, I.I. Maletina, N.V. Kondratenko and V.V. Orda, *Synthesis,* 1978, 835.
26. W.H. Huang, *J. Fluorine Chem.,* 1992, 58, 1.
27. N. Kamigata, T. Ohtsuka, T. Fukushima, M. Yoshida and T. Shimizu, *J. Chem. Soc. Perkin Trans. 1,* 1994, 1339.
28. W.R. Hasek, W.C. Smith and V.A. Engelhardt, *J. Am. Chem. Soc.,* 1960, 82, 543.
29. T. Kitazume and N. Ishikawa, *Chem. Lett.,* 1982, 137.
30. V.C.R. McLoughlin and J. Thrower, *Tetrahedron,* 1969, 25, 5921.
31. G.J. Chen and C. Tamborski, *J. Fluorine Chem.,* 1989, 43, 207.
32. G.J. Chen, L.S. Chen and K.C. Eaper, *J. Fluorine Chem.,* 1993, 63, 113.
33. G.J. Chen, *J. Fluorine Chem.,* 1990, 46, 137.
34. J.W. Labadie and J.L. Hedrick, *Polym. Prep.,* 1990, 31, 344. 32.
35. S.Y. Kim and J.W. Labadie, *Polym. Prep.,* 1991, 32, 164.
36. F. Bellini and K. Sestanj, U.S. Patent 1986, 4,604,406.
37. D.J. Burton, D.M. Wiemers and J.C. Easdon, U.S. Patent 1986, 4,582,921.
38. G.E. Carr, R.D. Chambers, T.F. Holmes and D.G. Parker, *J. Chem. Soc. Perkin Trans. 1,* 1988, 921.
38a. J.N. Freskos, *Synth. Commun.,* 1988, 965.
39. D.R. Buckle, D.S. Eggleston, I.L. Pinto, D.G. Smith and J.M. Tedder, *Bioinorg. Med. Chem. Lett.,* 1992, 2, 1161.
40. A.E. Feiring, *J. Fluorine Chem.,* 1977, 10, 375; *J. Org. Chem.,* 1979, 44, 1252.
41. C.L. Liotta and H.P. Harris, *J. Am. Chem. Soc.,* 1974, 96, 2250.
42. N. Ishikawa, T. Kitazume, T. Yamazaki, Y. Mochida and T. Tatsuno, *Chem. Lett.,* 1981, 762.
43. J. Ichihara, T. Matsuo, T. Hanafusa and T. Ando, *J. Chem. Soc. Chem. Commun.,* 1986, 793.
44. J.H. Clark, A.J. Hyde and D.K. Smith, *J. Chem. Soc. Chem. Commun.,* 1986, 791.
45. T. Ando, D.G. Cork, M. Fujita, T. Kimura and T. Tatsuno, *Chem. Lett.,* 1988, 1877.
46. J. Ichihara, T. Hanafusa, Y. Takai and Y. Ito, *Chem. Lett.,* 1992, 1161.
47. H. Sonoda, T. Sonoda and H. Kobayashi, *Chem. Lett.,* 1985, 233.
48. M.A. McClinton and D.A. McClinton, *Tetrahedron,* 1992, 48, 6555.
49. Y. Kobayashi, K. Yamamoto and I. Kumadaki, *Tetrahedron Lett.,* 1979, 42, 4071.
50. A. Ando, T. Miki and I. Kumadaki, *J. Org. Chem.,* 1988, 53, 3637.
51. C.H. Depuy and A.L. Schultz, *J. Org. Chem.,* 1974, 39, 878.
52. B.R. Langlois, *J. Fluorine Chem.,* 1988, 41, 247.
53. J. Hine and J.J. Porter, *J. Am. Chem. Soc.,* 1957, 79, 5493.
54. J. Hine and K. Tanabe, *J. Am. Chem. Soc.,* 1958, 80, 3002.
55. T.G. Miller and J.W. Thanassi, *J. Org. Chem.,* 1960, 25, 2009.
56. A. Marhold, Paper 13, Fluorine in Agriculture, Manchester (UK), 1995.

Chapter 8

Industrial Aspects of Aromatic Fluorine Chemistry

8.1 INTRODUCTION

The development of organofluorine chemistry, including aromatic fluorine chemistry, has been closely associated with the need for speciality and effect chemicals and can be traced back to the extensive research in the 1940s largely directed towards the search for materials that were resistant to uranium hexafluoride needed for uranium isotope enrichment. The high cost of organofluorine compounds (due to the difficulties in synthesis and handling and not due to any global shortage of fluorine — see Chapter 1) precludes them from most large-scale applications and makes them better suited for the high-cost and low-turnover areas of chemical specialities.

Aromatic fluorine compounds are relative newcomers on the industrial scene compared to heavily fluorinated fluoroaliphatics such as those used in plastics, solvents, and propellants. Fluoroaromatics were prohibitively expensive until the 1960s when the "halex" (halogen exchange) method became available as a viable alternative to vapor phase perfluorination followed by rearomatization and selective removal of fluorine.[1] The success of the halex method and the effective development of diazotization methods (which have the potential for continuous processes, such as in the manufacture of fluorobenzene from aniline), coupled with a rapidly increasing level of commercial interest in higher value products, led to a rapid growth in commercial interest in aromatic fluorine compounds. Thirteen thousand patents were filed worldwide between 1981 and 1986 for fluorine-containing pharmaceutical and pesticide products and by the late 1980s there were established markets for aromatic fluorine compounds in several sectors, notably anti-inflammatories, antibiotics, fungicides, pesticides, herbicides, and in a few other areas, such as dyes. The commercial diversity of aromatic fluorine compounds has grown to include liquid crystal products and polymers.

8.2 THE DEMAND FOR AROMATIC FLUORINE COMPOUNDS

World consumption of aromatic fluorine compounds for all major applications is 4 to 4.5×10^3 tonnes per annum. The agrochemicals market is the largest consumer, at up to 2×10^3 tonnes per annum, followed by pharmaceuticals, at about 1.5×10^3 per annum, with the rest of the market including polymers, dyes, and liquid crystals. The total market is expected to grow at about 7% per annum between now and the turn of the century.

The agrochemicals market alone is estimated to be worth about $25 billion of final products (as bought by the farmer) per annum.[2] This actually represents a real decrease in the global market of about 5% and indeed the market has been falling over several years, due to the European Union's Common Agricultural Policy reforms, the uncertainties surrounding the GATT negotiations, and dramatic changes in the Eastern Block, including the former Soviet Union. The predictions for the immediate future are more optimistic, however, with real market growth expected as political reforms work their way through the system.

The value of basic aromatic fluorine intermediates in the agrochemicals sector is only $30 to $35 million per year, but still represents a significant increase over the last 10 years. The figure is deceptive, however, as it only represents a small fraction of the final sales value of the products. One particularly successful product, trifluralin (2,6-dinitro-4-trifluoromethyl-N,N-dipropylaniline), a herbicide, had sales that peaked at greater than $400 million before it was largely overtaken by generics (new fluorine-containing products from the same company, Dow Elanco, have more recently helped to recapture the market). Up until the early 1980s there were few fluorine-containing crop protection products on the market, with the best estimates[3] suggesting 40 products either in the marketplace or in late development stages (Trifluralin was already an established product with sales approaching their peak). About 40% of these had herbicidal activity, nine were insecticides, and six were fungicides, the rest being rodenticides or plant growth regulating materials. There are now some 70 products, of which 15 to 20 are in the marketplace. The majority of the new products are herbicides or fungicides. The number of products and total sales of aromatic fluorine-containing agrochemicals remains low, however, and this is due partly to the high contribution of the costs of active ingredients to the actual selling price of the finished product (20 to 25%), which weighs against relatively expensive fluorine products. The entry of Indian and Chinese producers of aromatic fluorine intermediates has helped to reduce the costs and this, coupled with the expected upturn in the overall market for agrochemicals, should improve these figures. However, the number of late development products is below that for pharmaceuticals, although it is still considerable, and several of the new products are likely to become significant in the next 5 to 10 years. These are most likely to be herbicides. It is encouraging to note that in a recent review of the important 8th IUPAC Congress of Pesticide Chemistry (which attracted over 2100 delegates), 9 of the 27 highly promising new products described contained fluorine.[4]

Of the total worldwide market for pharmaceuticals of some $175 billion, a remarkable $10+ billion is due to fluorine-containing products. This is based on some 500 fluorine-containing products out of a total of 10,000 products. The percentage of the sales costs due to the costs of the active ingredients is somewhat lower than that for agrochemicals, being about 10%. The value of the fluorinated intermediates contributing to these fluorine-containing products is about $100 million — some three times larger than that for agrochemicals, despite the larger volume turnover of the latter. Over half of this market is due to a few blockbusters. At present the leading F-drug is Cipro (Bayer), an antibiotic with total sales of $1.2 billion (compare the leading drug Zantac at $3.6 billion). Other major F-drugs are fluoxetine (Lilley), an antidepressant ($1 billion); fluticasone (Glaxo), a fluorinated steroid ($1 billion); and flucanazole (Pfizer), an antifungal ($0.65 billion). Likely major F-drugs for the future are the asthma-steroid fluticasone and the anti-ulcer agent lansoprazole. It is probable that sales of F-drugs will continue to grow despite the tightening of the worldwide pharmaceutical market due to stricter financial controls. It is estimated that the total market for fluorine-containing drugs will increase to about $15 billion by the turn of the century, representing an increase of 7% per annum, twice the maximum likely growth rate for pharmaceuticals overall. It is likely that some $10 billion of this will be due to about 10 blockbusters.

The world sales of fluorine-containing reactive dyes in 1990 was estimated at 10,000 tonnes,[5-10] representing about 20% of the market and an impressive growth from the first fluorine-containing dye (5-chloro-2,6-difluoro-4-pyrimidinyl)[11] launched in 1966 by Bayer and Sandoz. The reactive dye business is undergoing radical changes, however, through the impact of ecological demands and environmental legislation. Fluorine promises to continue to play a major role in the development of new, environmentally friendly systems and some of the new bifunctional dyes do indeed contain fluorine.[11] Other application area for dyes have seen relatively little success, with only a few azo dyes containing fluorine, for example, and those almost exclusively contain the CF_3 group.

The effect and advantages of fluorine substitution on lower-molecular-weight liquid crystals has been well known since the early 1980s and this has led to the rapid development of various high value but low turn-over products often incorporating on F-, CF_3-, and CF_3O-substituted benzenes. The rather less straightforward effects of fluorine on multicore products is a drawback, however.[12] The future demand for fluorine-containing aromatics in this area is likely to be significant but relatively small.

Other potential growth areas for fluorine-containing aromatics are still less certain, although it seems likely from developments in the last 20 years that they will emerge. Fluorine already has an established place in aliphatic polymer chemistry and new materials in this category will continue to find applications in rather specialist areas.[13] Fluorinated aliphatic groups bridging aromatic rings (especially via the use of hexafluoroacetone as a bridging group) in polymers[14] are gaining popularity as possible advanced materials, as are actual ring-fluorinated polyaromatics[15-17] with short- to medium-term possibilities in sectors such as sporting goods, high-performance cars, electronic circuit board coatings, aerospace applications, and, in the longer term, the enormous automobile market. These could prove to represent as important an outlet for fluorine-containing aromatics in the future as pharmaceutical products.

8.3 THE SUPPLY OF AROMATIC FLUORINE COMPOUNDS

The major manufacturers of fluoroaromatics are listed in Table 8.1. There are also several minor manufacturers of fluoroaromatics, including Solvay Duphar (Netherlands), Tokkem (Japan), Ihara (Japan), Wychem (UK), and an increasing number of Indian and Chinese producers. The total world capacity for production of fluoroaromatic intermediates in 1995 is about 7500 tonnes/year, with about 1500 tonnes/year being for internal consumption. There is, therefore, an over capacity of something

Table 8.1 Leading Fluoroaromatic Manufacturers

Company	Approximate Capacity (tonnes/year)
Miteni (Italy)	1000
Zeneca (UK)	2000–2500
Asahi Glass (Japan)	1000
Hoechst (Germany)	800
Rhone Poulenc (France)	800
Riedel de Haen (Germany)	800

in excess of 1000 tonnes/year, which should be taken up by the end of the century if the expected market growth is realized.

The production of fluoroaromatics provides an interesting example of the rationalization that has affected the chemical industry in recent years. In 1990, nine companies in Western Europe alone produced fluoroaromatics on a significant scale. Perhaps the single most important change in that market has been the withdrawal of Shell chemicals following the major accident at their manufacturing site in Stanlow (UK).

Halex (halogen exchange) continues to be the most popular production method being used by Asahi Glass, Hoechst, Miteni, and Rhone-Poulenc, among others. The hazards associated with the isolation of diazonium intermediates in diazotization-based methods has to some extent been overcome by the successful development of continuous HF-diazotization technology, most notably that developed by Zeneca at their Grangemouth site, which is probably the largest fluoroaromatic plant in the world. Riedel de Hahn continue to enjoy the reputation for being the world leaders in innovative fluorinated aromatic products and in their flexibility of processes and products.

There are relatively few side-chain fluorinated aromatic compounds in production, despite the large overall size of the market and the very large number of products in development. The technology required is very similar to that employed in aliphatic fluorination, so that the leaders in this sector are typically significant players in the manufacture of side-chain fluorinated aromatics. Rhone-Poulenc have been the largest producers of side-chain fluorinated aromatics in Europe in the 1990s, with other significant manufacturers including Hoechst, Miteni, and Dow-Elanco (France).

8.4 INDUSTRIAL METHODS

Aspects of the manufacturing processes based on halex (Chapters 2, 5, 6, and 7) and diazotization (Chapter 3) routes to selectively fluorinated aromatic compounds are briefly considered below. The first industrial report of a halex process for the manufacture of a fluoroaromatic came from the U.S. in 1936 and involved the solvent-free preparation of 1-fluoro-2,4-dinitrobenzene from the chloro-analogue using KF at high temperatures.[13] Ten years later, the halex process was improved by the introduction of a solvent, and most significantly, after a further 10 years, the first description of a KF/dipolar aprotic solvent reaction system effectively opened the door to full commercialization of halex chemistry.

The first industrial reports on the analogous halex-type process (but involving acidic reagents, i.e., anhydrous HF or anhydrous $HF/SbCl_5$) for converting CCl_3 groups attached to aromatic nuclei to CF_3 groups also appeared in the 1930s, and manufacturing processes were running in Germany by the end of that decade.[13]

In modern halex manufacturing processes for ring fluorination, a major factor that needs to be taken into consideration is corrosion (as is the case with all fluorination processes and even nonfluorination processes that involve the handling of fluorocompounds). For this reason, reactions are normally run in a stirred heated vessel made out of a fluoride-resistant material such as the nickel-based alloy Inconel. The fluoride (typically KF and usually based on a significant excess of F^- based on that required by the reaction stoichiometry and allowing for deactivation of some of the fluoride, possibly by poisoning by the inorganic chloride product crystals) and solvent (typically dipolar aprotic) are normally first dried together by heating to remove water, often via azeotropic drying with a convenient inert and inexpensive solvent such as toluene (which can be recycled). Once dry, the substrate and, if required, phase transfer catalyst are added and the reaction mixture is taken up to temperature for the period of time required (commonly several hours).

The high cost of the solvent used in halex processes and environmental concerns about the use and recovery of solvent, makes the solvent a very important factor in process design. The quantity of solvent may be kept down to the minimum required to completely dissolve the substrate and fluidize the heterogeneous mixture. Separation and product isolation must involve efficient solvent recovery methods that would normally not involve such a water quench. The solvent and product would preferably be distilled away from the inorganics. There is a growing trend, however, towards the development of no-solvent or very low solvent processes where the process engineering may be significantly different.

The Balz-Schiemann reaction, first reported in the literature in 1927, was the basis of commercial processes for the manufacture of fluoroaromatics by the second World War.[13] The plant procedure is made complex by the need for two separate reactions. The initial diazotization requires a cooled vessel with good agitation. The aniline would normally be dissolved in excess dilute HCl followed by addition of a solution of sodium nitrite at about 10°C. Excess fluoroboric acid solution is added on completion of the diazotization, leading to precipitation of the $ArN_2^+BF_4^-$, which can be recovered by filtration, washed (it is important to remove chlorides, which can otherwise lead to chloroaromatic side products in the final stage of the reaction), and thoroughly (but cautiously) dried. Most of the aryl tetrafluoroborates are quite stable but care is needed in their handling. They are generally severe irritants and those possessing electron-withdrawing substituents on the ring(s) may actually not be stable enough for isolation.

The final stage of the process is the dediazotization, which will require heating either the dried salt or a slurry of the salt and an inert medium such as sand or paraffin. Commonly, temperatures can be as high as 200°C but care must be exercised to avoid runaway reactions and a method whereby only small amounts of solid are heated at a time is required. The fluoroaromatic is fractionated from the thermally treated salt, and the liberated BF_3 trapped in aqueous HF, enabling recycling of the tetrafluoroboric acid.

The 1930s also saw the first industrial reports on the "one-pot" diazotization method for the production of fluoroaromatics using the aniline/sodium nitrite/HF process first reported in the previous century. A manufacturing plant based on this chemistry for the preparation of fluorobenzene was operational in Germany by the 1940s.[13] The great potential advantage of this method over the Balz-Schiemann method is the use of an *in situ* decomposition of the intermediate diazonium fluoride. There is also no need for gaseous recycling, as the only side product is nitrogen. One of the limitations of the process is the low boiling point of the HF system (HF is used as reagent and solvent), which can be below the temperature necessary to affect the decomposition of many substituted aryl diazonium fluorides (notably those containing electron-withdrawing substituents). Water may be added to increase the boiling point, although this may lead to phenolic by-products.

Reaction vessels for the HF diazotizations must be made of special materials that can resist the highly corrosive acids. Originally this led to regular replacement of the reactors, which obviously added to the process costs. Nowadays, improved alloys and the availability of HF-resistant plastic-lined vessels have largely overcome this problem, although the corrosive nature of HF as well as the unique burns that it can cause to human skin are significant concerns to plant operators. Normally, the HF is added to the vessel along with any water and NaF to provide the right acid strength/boiling point. The aniline is then added to the cooled HF followed by the sodium nitrite. The mixture is then heated under reflux, normally in a separate vessel, after removal of excess NOF (the reactive intermediate):

$$NaNO_2 + 2HF \rightarrow NOF + H_2O + NaF$$

$$ArNH_2 + NOF \rightarrow ArN_2^+ F^- + H_2O$$

$$ArN_2^+F^- \rightarrow ArF + N_2$$

It is important that the HF, which is used in considerable excess, is recycled. The process is exothermic at all stages and care must be taken to control these exotherms.

The direct HF-based diazotization route to fluoroaromatics is now an important manufacturing process that can be made continuous. It rivals the halex process in terms of volume of products produced and surpasses the more familiar Balz-Schiemann method, which is preferred only in small-scale manufacturing.

8.5 INDUSTRIAL PRODUCTS

The major agrochemical products that consume aromatic fluorine compounds in their manufacture are used in all the main segments of the industry, as herbicides, fungicides, and as insecticides.[18-23] The established industrial products are based on fluoroaromatics and on benzotrifluorides, with some containing fluorinated hetero-cyclic units. Of the 20 or so commercially significant compounds in this class, over half of these are herbicides. The best known of these is a trifluoromethylated aro-matic, trifluralin (Figure 8.1), which has been in the market for about 30 years and has become a commodity with peak sales of about 6000 tonnes. As this product was

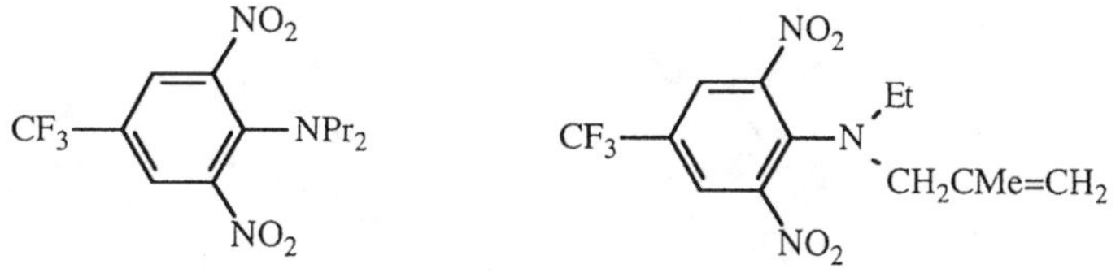

Trifluralin

FIGURE 8.1   Dinitroaromatic agrochemicals.

peaking in sales in the early to mid-1980s, derivatives were developed, including ethalfluralin (Figure 8.1). Some of the other established products are given in Figure 8.2 so as to illustrate the overall structural diversity of the products, but also to illustrate the common occurrence of ring F and CF_3 groups.

Structure-activity relationships on agrochemicals can help to illustrate the value of fluorine-containing substituents and their relative positions. Thus, the postemergent herbicidal activities of 2,6-diarylpyrimidines with different substituents (Figure 8.3) show enhanced activity with a combination of a meta substituent on one ring and a para substituent on the other ring. By keeping one substituent as CF_3 the relative activity of substituents on the other ring clearly indicates the benefits of fluorine with ring F and ring CF_3 being preferred.[21]

Among the major herbicide innovations of the 1980s were the sulfonylureas and aromatic fluorine compounds are again playing a major role in their development. An example of a fluorine-containing sulfonylurea herbicide is shown in Figure 8.4. Compounds of this type are very active while also being highly selective in their action on maize, so that on postemergent application they only kill the weeds, whereas nonfluorinated analogues will also attack the maize.

There are many agrochemicals that contain aromatic fluorine compounds at development stage, as witnessed by the very high level of patent activity which has been maintained for over 10 years (about 20,000 patents on fluorine-containing agrochemicals and pharmaceuticals in the last 10 years). The trifluoromethyl group again figures prominently, as does aromatic fluorine. What is equally significant, however, is the emergence of new fluorine-containing aromatic substituents such as OCF_3 and the significant number of products that contain fluorine-containing groups on heterocyclic substituents. Some of the many development agrochemicals that have been recently reported are shown in Figure 8.5.[4,24-27]

Fluorine-containing medicinals feature significantly in many areas, including anticancer and antiviral drugs, anti-inflammatory and antiparasitic agents, antibiotics, anesthetics, and the treatment of cardiac and mental illnesses, as well as in artificial blood substitutes and medical diagnosis (e.g., X-ray contrast agents).[28] In several of these areas, notably in drug design, aromatic fluorine compounds feature significantly and are the subject of an enormous patent literature and associated commercial research and development effort.

One of the best known of all the fluorine-containing medicinals is 5-fluorouracil (Figure 8.6), which has significant tumor-inhibiting activity.[29] 5-Fluoro-2'-deoxy-β-uridine (Figure 8.6) is now known to be more effective and less toxic than

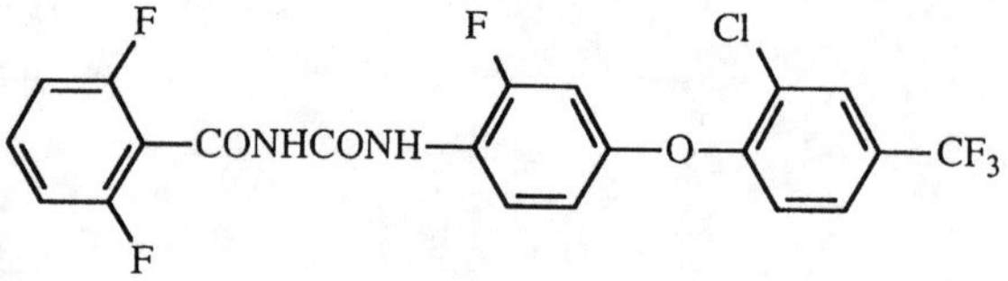

Flufenoxuran (larvacide)

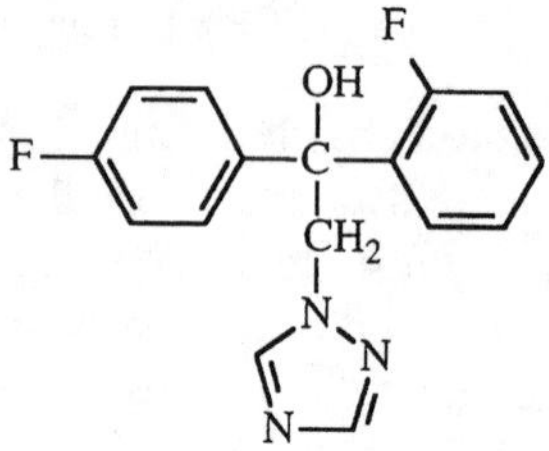

Flutriafol (systemic fungicide)

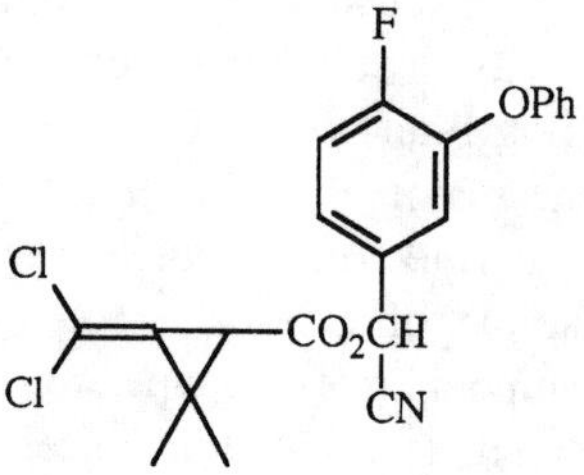

Cyfluthrin (synthetic pyrethroid)

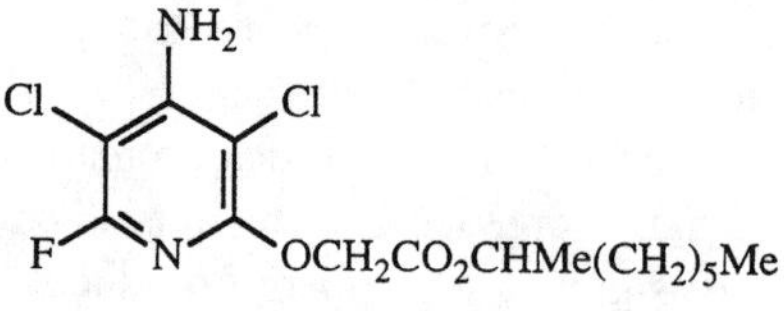

FIGURE 8.2 Some established agrochemicals based on aromatic fluorine compounds.

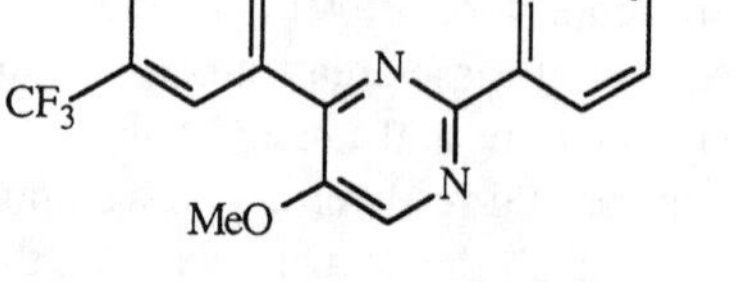

Fluroxypyr (post-emergence herbicide)

FIGURE 8.3 Relative postemergent herbicidal activities of diarylpyrimidines.

R = F > CF$_3$ > Cl > H

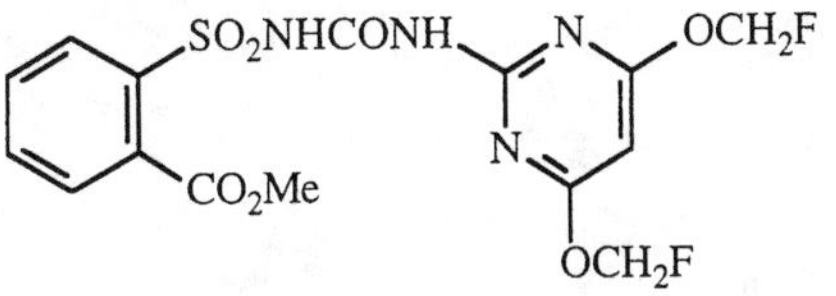

FIGURE 8.4 A postemergent herbicide with high specificity towards weeds.

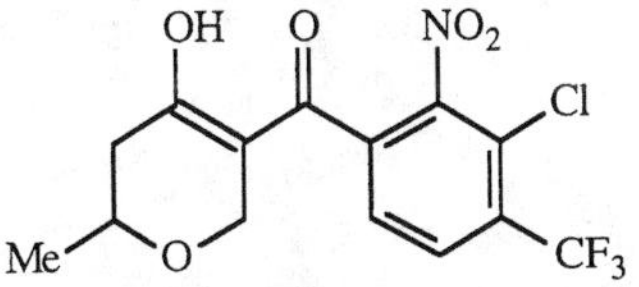

Primisulfuron

Benzoylpyrone herbicide

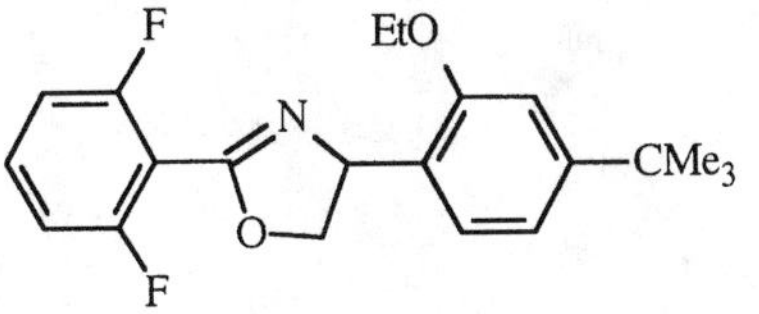

Oxazoline insecticide

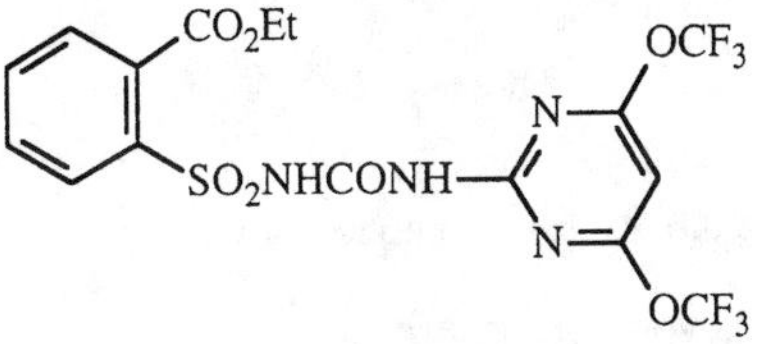

FIGURE 8.5 Some agrochemicals in development.

Sulfonylurea herbicide

5-fluorouracil. These compounds are anabolized to a potent inhibitor of DNA synthesis in the tumor cells (where they concentrate) as a result of the presence of the unreactive fluorine. Further elaborations on these anticancer agents, including alternative active components (Figure 8.6) and formulations, are in development.[28]

Other aromatic fluorine compounds have been developed as anticancer drugs and are undergoing trials. These include fludarabine phosphate (Figure 8.7), a purine antimetabolite which is resistant to adenosine deaminase, and a quinolinecarboxylic acid (Figure 8.7), which shows broad-spectrum *in vitro* and *in vivo* activities towards a variety of human and murine tumors.

While fluorine-containing steroids have been important medicinal products since the 1950s, the undesirable side effects of steroid drugs has led to the development of nonsteroidal anti-inflammatory drugs which would be more active than the salicylates and effective in the treatment of conditions such as rheumatoid arthritis.

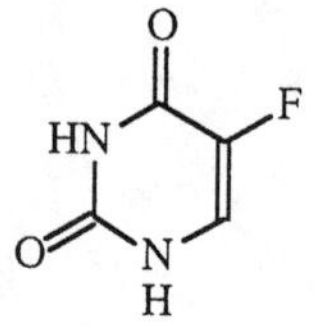

5-fluorouracil

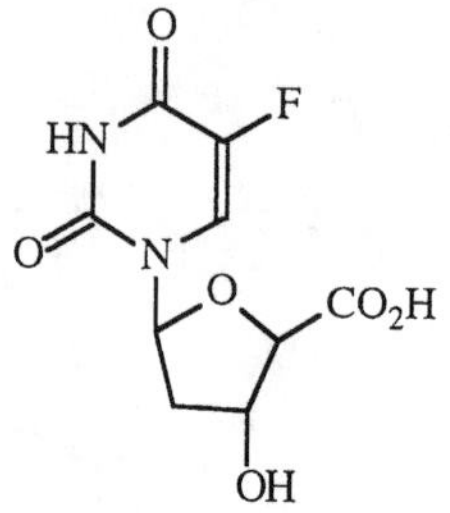

5-fluoro-2'-deoxy-β-uridine

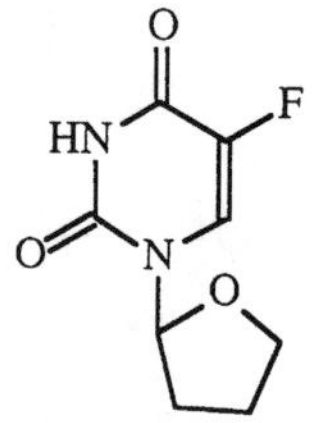

N,-tetrahydrofuranyl-5-fluorouracil
(futraful)

FIGURE 8.6 Anticancer agents based on 5-fluorouracil.

Several fluoroaromatic products have proven to be very successful. These include diflunisal (traded under the name Dolobid by MSD and recommended for the treatment of rheumatoid arthritis and osteoarthritis), flurbiprofen (traded under the name Froben by Boots and recommended for the treatment of rheumatoid arthritis, osteoarthritis, and acute muscoskeletal disorders), and sulindac (traded under the name of Clinoril by MSD and recommended for the treatment of various forms of arthritis and periarticular disorders) (Figure 8.8). Aromatic fluorine anti-inflammatory drugs in development stages include those based on imadazole.[27]

Several antibiotics based on aromatic fluorine compounds with CF_3 and F groups are in use, including mefloquine (Figure 8.9), which is one of the most effective antimalarial drugs, 5-fluorocytosine (Figure 8.9), which is an antifungal agent, and flucloxacillin (Figure 8.9), a narrow-spectrum antibiotic. Other important products include the fluoroaromatic fluconazole, which has excellent activity in a remarkably wide range of fungal infections, and ciprofloxacin, a wide-usage product that is one of the market leaders in the quinolone business. There are in fact several fluorine-

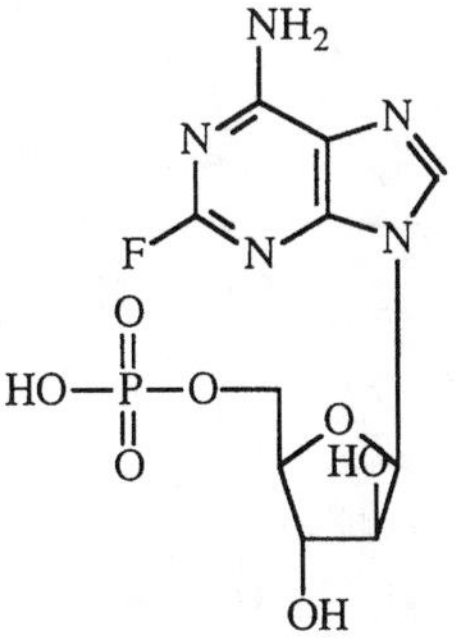

Fludarabine phosphate

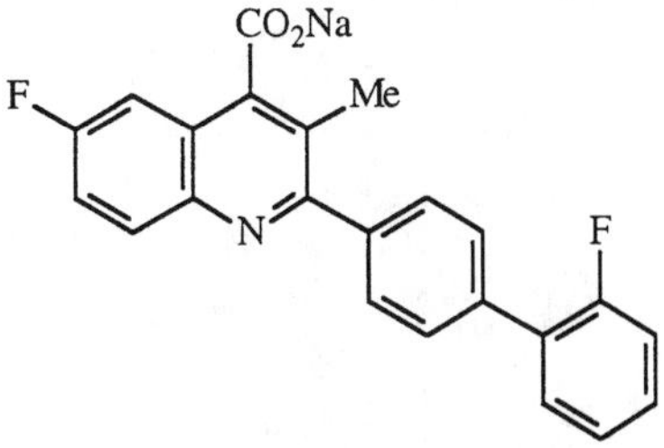

FIGURE 8.7 Some fluorinated development-stage anticancer drugs.

DuP-785

containing quinolones on the market, although one of them, temafloxacin had to be withdrawn after just a few weeks due to serious side effects. Several new fluoroquinolones are heading for the market, including fleroxacine and lomefloxacin, which are again based on fluoroaromatic building blocks.[26]

The trifluoromethylated benzothiadiazine, bendroflumethiazide (Figure 8.10) is one of the most powerful of several fluorinated compounds used as diuretics and antihypertensive agents.

The importance of fluorine and fluorinated groups such as CF_3 in increasing the lipid solubility of drugs is clearly illustrated in the value of many fluorine-containing central nervous system agents. Several aromatic fluorine-containing neuroleptics, antidepressants, sedatives, muscle relaxants, and other CNS agents are in use and their very high activity is likely to be at least partly due to the rate of absorption and transport of the drugs across the blood-brain barrier (Figure 8.11). A wide variety of structural types are again evident with aromatic fluorine and trifluoromethyl groups featuring prominently.

Despite the large number of patented and published reactive components for dyes that contain fluorine, only three have attained any real economic significance (Figure 8.12). These form the basis for several dyestuffs where one or two of the fluorines have been replaced by amino groups. These useful intermediates are six-membered heterocycles and alternatives such as polyfluorobenzenes and fluorinated five-membered heterocycles are of no commercial value.[30] While the fluoropyrimidines

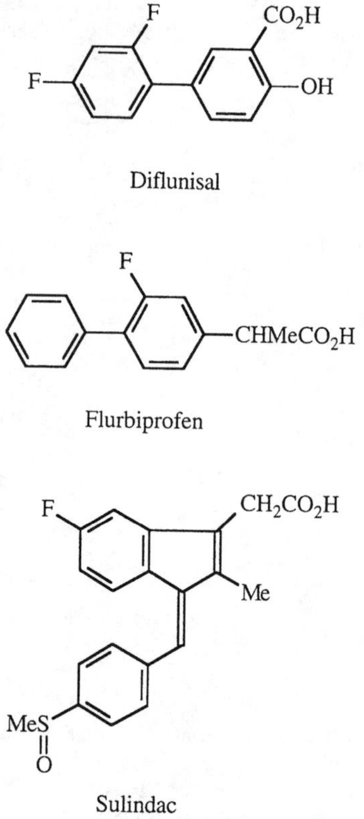

Diflunisal

Flurbiprofen

FIGURE 8.8 Fluorine-containing nonsteroidal anti-inflammatory drugs.

Sulindac

and triazines (Figure 8.12) are proven products, fluorinated pyridines have limited use in reactive dye manufacture. Fluorinated pyridazinones may prove to be more reactive and should produce dyeings with comparable fastness properties to the commercially useful chlorinated analogues. There is also considerable interest in fluorinated dyestuffs (i.e., where the fluorine remains in the dye after application) notably CF_3-substituted azo dyes,[31] although these are as yet unproven commercially. Fluorine-containing substituents lower the energy of the intermolecular interactions and the heat of evaporation of dispersive azo dyes by 10 to 40 kJ mol^{-1} compared with the unsubstituted dyes, and can therefore be useful for heat-transfer printing of fabrics. Some examples of these product types are shown in Figure 8.13.

The ability of fluorine to exert a large influence on permittivity while causing minimum change in molecular shape (strong electron-withdrawing effect and small size) makes it a likely substituent in the design of liquid crystals.[32] There are several different sites for substitution in most liquid crystals; usually fluorine or a small fluorine-containing substituent such as CF_3 or OCF_3 is placed at a terminal position, or fluorine itself can be located on the structural rings (less commonly, the bulkier fluorinated groups). Product types include Schiff bases, benzoates, biphenyls, and terphenyls (Figure 8.14). In general some of these products have proven to be useful by themselves or in formulations, including the more conventional cyano-terminated products.[12]

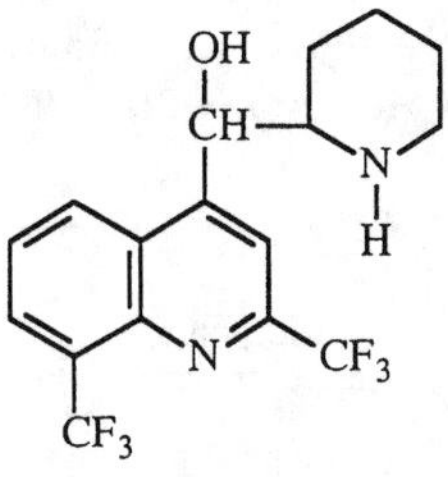

Mefloquine

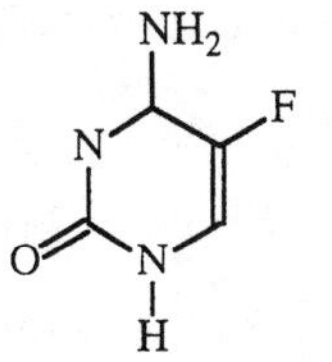

5-Fluorocytesine
(flucytosine)

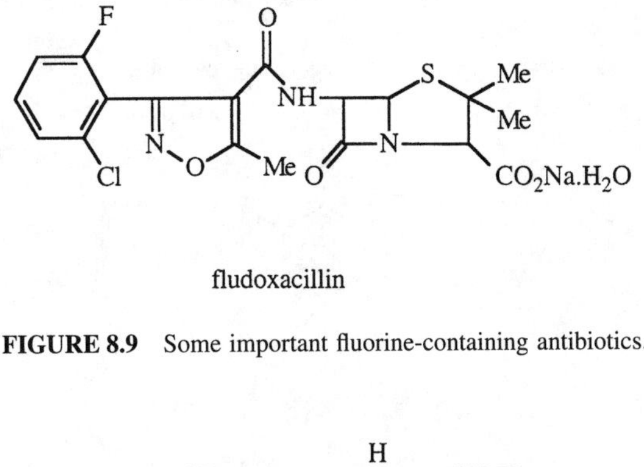

fludoxacillin

FIGURE 8.9 Some important fluorine-containing antibiotics.

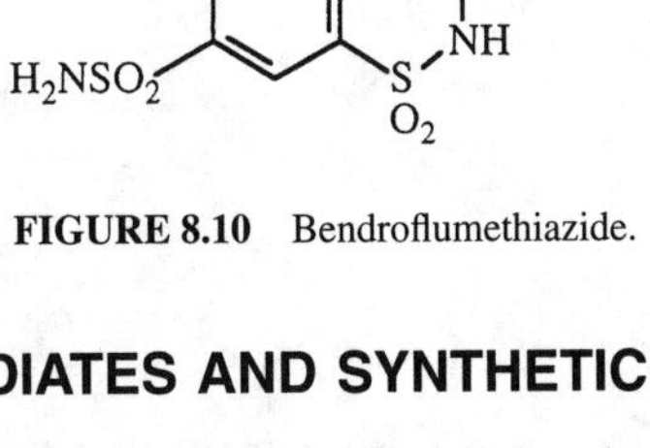

FIGURE 8.10 Bendroflumethiazide.

8.6 KEY INTERMEDIATES AND SYNTHETIC ROUTES

There are a relatively small number of key aromatic fluorine compound intermediates used in the manufacture of agrochemicals. These include fluoroaromatics such as 4-fluoro-3-hydroxybenzaldehyde, 2,6-difluorobenzamide, and fluorobenzene itself;

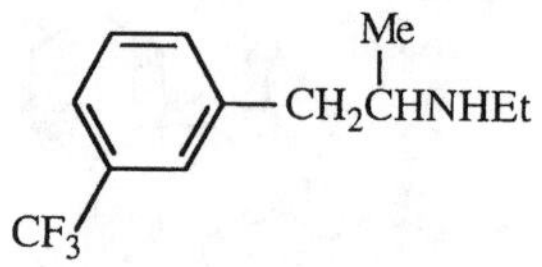

Fenfluramine
(appetite depressant)

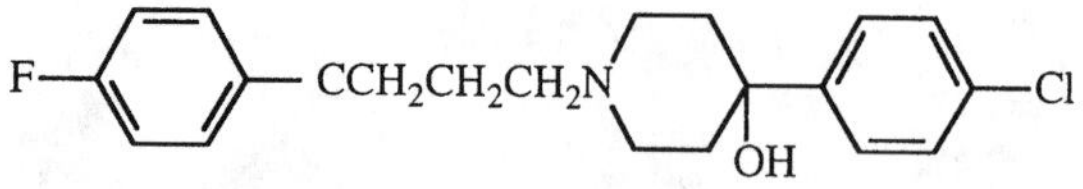

Haloperidol (management of schizophrenia)

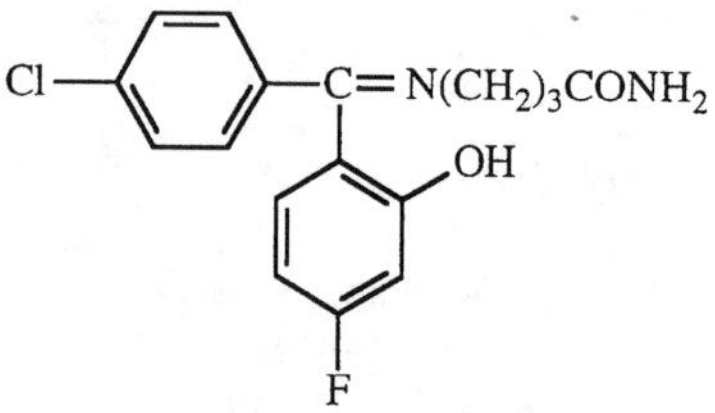

Progabide (antiepileptic drug)

FIGURE 8.11 Examples of fluorine-containing central nervous system agents.

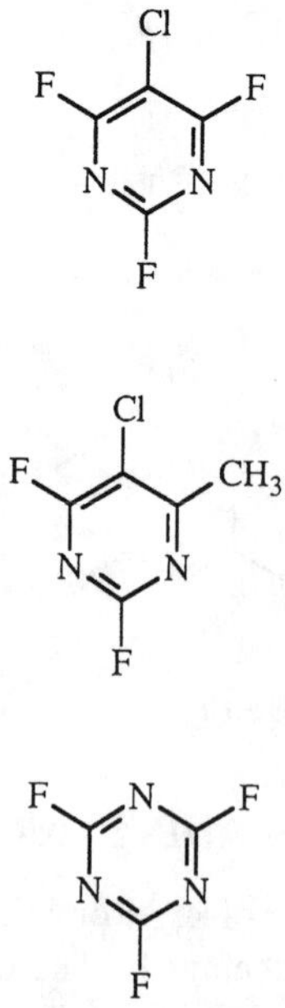

FIGURE 8.12 Commercially important fluorine-containing components for reactive dyes.

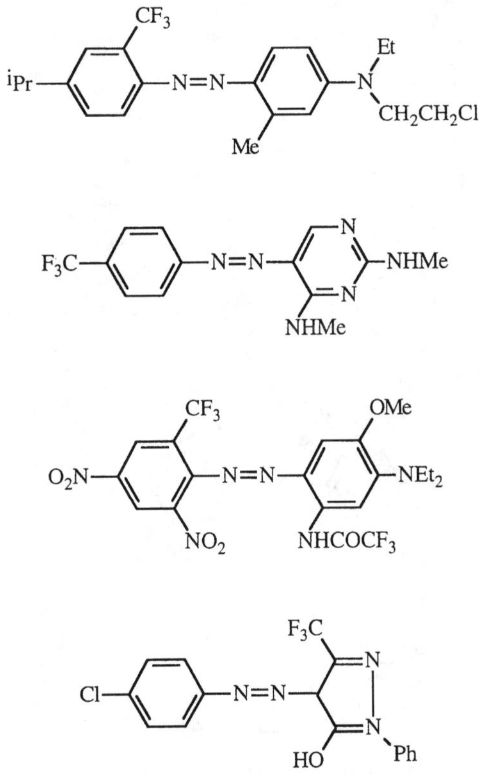

FIGURE 8.13   Fluorine-containing azo dyes.

benzotrifluorides such as 4-chlorobenzotrifluoride and 3-aminobenzotrifluoride; and trifluoromethoxyaromatics such as trifluoromethoxybenzene. A more comprehensive listing is given in Table 8.2. Some of the key aromatic fluorine compounds used as intermediates in the synthesis of pharmaceuticals are given in Table 8.3.

Common fluoroaromatic structural units in agrochemicals and pharmaceuticals include benzene groups with one fluorine, two meta fluorines, and, less commonly, two ortho fluorines, with occasional reports of interesting products based on benzene units with three or even four fluorines. Halex technology plays as important a role as any method in synthesizing aromatic fluorine compounds as intermediates for commercial and development products. The insecticide diflubenzuron is synthesized from the 2,6-dichlorobenzonitrile by reaction with KF (Figure 8.15) en route to the important intermediate 2,6-difluorobenzamide (Table 8.2). The successful application of diflubenzuron has led to the development of derivatives based on the same intermediate, including teflubenzuron, flufenoxuron, and chlorofluazon (Figure 8.15), all of which involve additional fluorine (F or CF_3) functions in the molecule.[23] Polysubstituted aromatic functions may be synthesized via an initial perchlorination stage. Thus, the herbicide fluroxypyr is prepared starting from pyridine which is perchlorinated followed by partial halex fluorination and then introduction of the other substituents (Figure 8.15). It is interesting to note that only two of the original

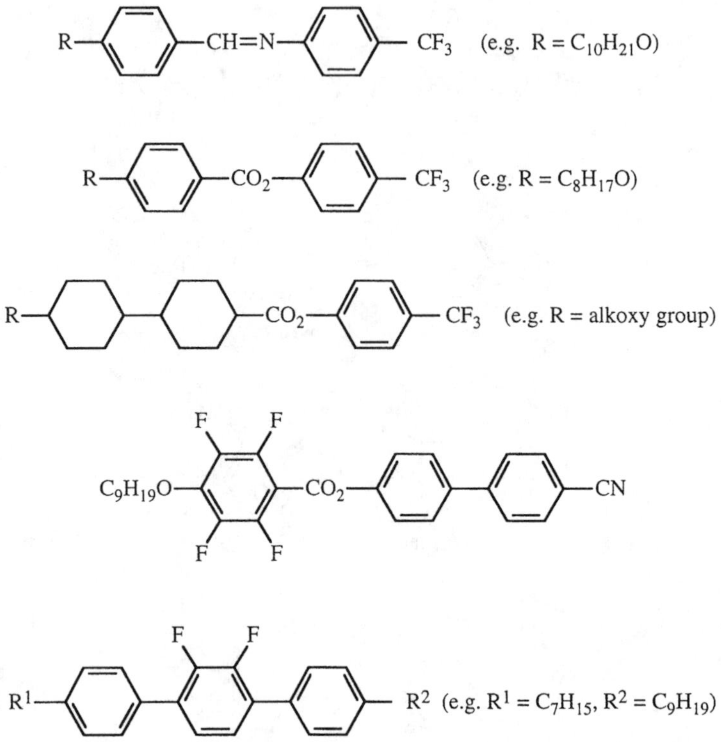

FIGURE 8.14 Examples of fluorine-containing liquid crystals.

chlorines introduced remain in the final product and, perhaps surprisingly, only one out of three fluorines. This is despite the high cost of fluorine in organic molecules, which can of course be an excellent leaving group. Loss of two out of three fluorines initially introduced in the synthetic pathway also applies in the synthesis of the fluoroquinolone fleroxacine, which employs 2,3,4,5-tetrafluorobenzoic acid as the key intermediate.[26]

4-Fluorophenol is a valuable intermediate in the synthesis of numerous agrochemicals and pharmaceuticals (Tables 8.2 and 8.3). It is interesting to note that there are some nine possible routes to this intermediate based on starting materials including fluorobenzene (now readily available as a result of the new Zeneca continuous diazotization process), phenol (involving direct fluorination, which is unlikely to be of commercial interest in all but highly specialized cases), and 4-fluoroaniline (an obvious if expensive route requiring the 4-fluoronitrobenzene intermediate, which can be prepared via halex technology). Generally, the routes can be broken down into those where the fluorine is introduced at an early or late stage in the synthesis. The currently preferred routes are via 4-fluoroaniline and 4-bromofluorobenzene, both of which are synthesized using the same technology of nitration, chlorination, fluorodediazotization, and reduction. These are, however, multistep processes and while alternatives such as those starting from phenol or

Table 8.2 Important Aromatic Fluorine Agrochemical Intermediates

Compound	Some Applications
Fluorobenzene	Flusilazole
	Flutrifol
	Epoxiconazole
	Nuarimol
4-Fluorophenol	Triazole fungicides,
	4-phenoxyquinoline fungicides
4-Chlorobenzotrifluoride	Trifluralin and related products
2,6-Difluorobenzamide	Chlorfluazuron
	Diflubenzuron
	Flufenoxuron
3-Chloro-4-fluoroaniline	Flampropmethyl
	Silafluofen
3,5-Dichloro-2,4-difluoroaniline	Teflubenzuron
3,4-Dichlorobenzotrifluoride	Various diphenylether products
2,6-Difluoroaniline	Diflufenican
Trifluoromethoxybenzene	Nuarimol
4-(Trifluoromethoxy)aniline	Dimilin-like benzoyl ureas and thioureas
3,5-Dichloro-2,4-difluoroaniline	Teflubenzuron
2-Fluoro-4-hydroxyaniline	Flufenoxuron
4-Fluoro-3-hydroxybenzaldehyde	Cyfluthrin
2,3,5,6-Tetrafluoro-4-methylphenylcarbinol	Tefluthrin
3-Aminobenzotrifluoride	Fluometuron

Table 8.3 Some Important Aromatic Fluorine Intermediates

Compound	Some Applications
4-Fluorophenol	Cisapriole
	Sorbinil
	Progabiole
	Sabeluzole
1,3-Dichloro-4-fluorobenzene	Ciprofloxacin
4-Fluoroaniline	Flumequine
3-Chloro-4-fluoroaniline	Norfloxacin
2,3-Difluoro-6-nitrophenol	Ofloxacin
2,3,4,5-Tetrafluorobenzoic acid	Fleroxacine
2,3,4-Trifluoroaniline	Lomefloxacin
2-Chloro-4-fluorobenzoic acid	Flosequinan

fluorobenzene offer the appeal of fewer steps their poor selectivity and low yields are major drawbacks (Figure 8.16).[31] It would seem that there is still considerable room for improvement in the synthetic routes to such simple intermediates as 4-fluorophenol. Other fluorinated phenol intermediates are useful intermediates in commercial processes. 3,4-Difluorophenyl esters are used in liquid crystals and can be synthesized by starting from 1,2-difluorobenzene, which is first nitrated, reduced to the aniline, and then converted to the phenol via diazotization — this is the preferred route.[12]

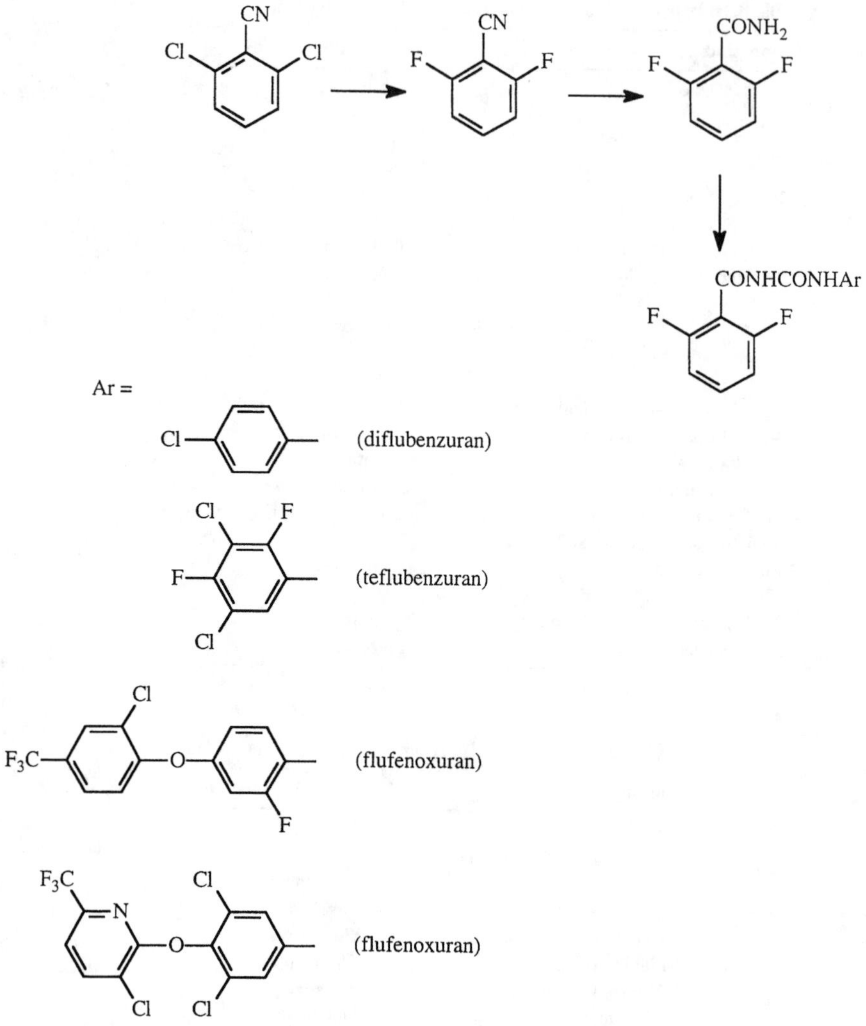
CN
Cl
Cl
CN
F
F
CONH2
F
F
CONHCONHAr
F
F
Ar =
Cl
(diflubenzuran)
Cl
F
F
Cl
(teflubenzuran)
Cl
F3C
O
F
(flufenoxuran)
F3C
N
Cl
O
Cl
Cl
(flufenoxuran)

FIGURE 8.15 Routes to insecticides involving halex chemistry.

Important benzotrifluoride intermediates generally contain N-derivatives (e.g., amino or nitro functions) or are themselves subject to nitration en route to the N-derivatives. The long established herbicide Trifluralin is synthesized in a four-stage process starting from 4-chlorotoluene and via 4-chlorobenzotrifluoride (Figure 8.17). The 3-aminobenzotrifluoride-based herbicide fluometuron is based on a more complex synthetic procedure due to the requirement of the CF_3 group in a meta position. The route starts from toluene to benzotrichloride, then to benzotrifluoride and on to 3-nitrobenzotrifluoride en route to the desired 3-aminobenzotrifluoride, which can then be reacted on to the product. In the synthesis of the 3-aminobenzotrifluoride, the isomer 2-aminobenzotrifluoride is also

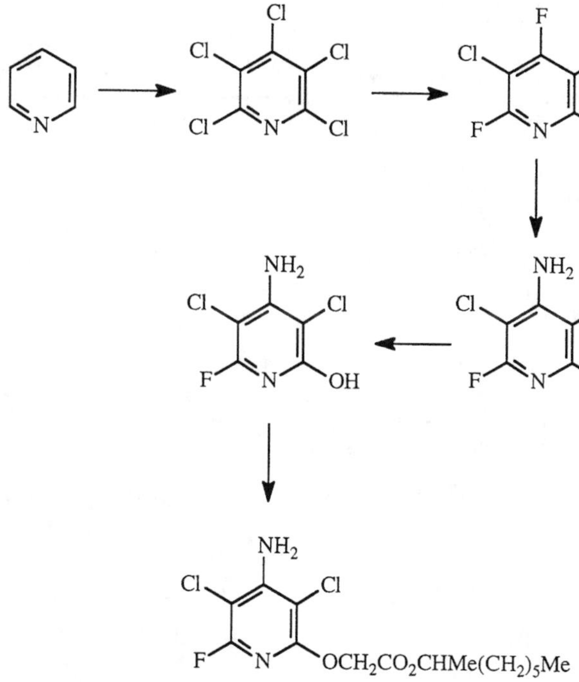

Fluoroxypyr

FIGURE 8.15 (continued).

produced (as a result of unselective nitration in the previous step) in small quantities and is a valuable intermediate in the synthesis of the antimalarial drug mefloquine.[26] The other isomer, 4-fluoroaniline is also produced in small amounts, but these quantities will be insufficient to meet the likely demand resulting from the new development of pharmaceutical and agrochemical products based on the 4-N-substituted benzotrifluoride unit.[26] New important benzotrifluoride intermediates include 5-amino-2-nitrobenzotrifluoride (for the pharmaceutical flutamide), its reduction product 2,5-diaminobenzotrifluoride (which can be used in the synthesis of new engineering polymers), 2-amino-5-chlorobenzotrifluoride (an intermediate in the synthesis of the fungicide, triflumizole), 4-trifluoromethylbenzaldehyde (for the synthesis of the insecticide hydromethylnon), 2-trifluoromethylbenzoylchloride (for the fungicide flutoanil), and 4-trifluoromethylbenzonitrile (for possible new herbicides). Rarely, organometallic routes are considered commercially for the synthesis of particularly valuable benzotrifluoride intermediates such as those used in liquid crystals (although the method may be more commonly used in the future). Thus, 4-hydroxy-4'-(trifluoromethyl)biphenyl can be prepared by first protecting the 4-hydroxy-4'-bromobiphenyl substrate by methylating the hydroxy group, then reacting the methoxy compound with Cu/CF$_3$I in DMF followed by a final deprotection of the hydroxy group with (CH$_3$)$_3$SiI.[12]

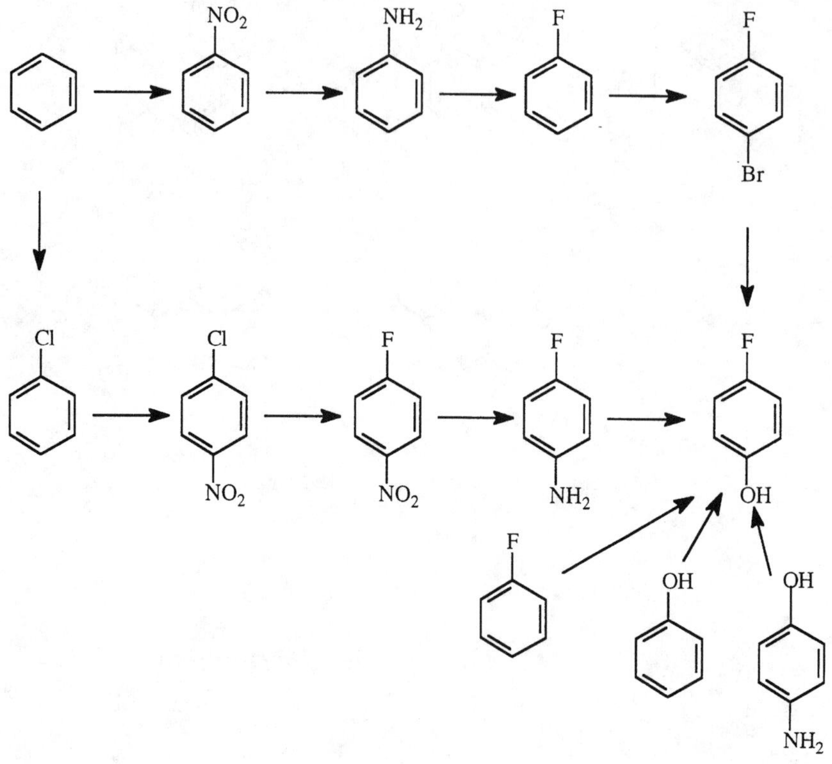

FIGURE 8.16 Some routes to 4-fluorophenol.

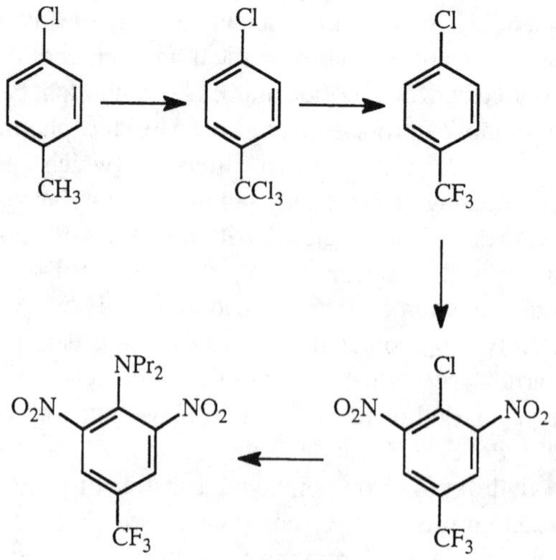

FIGURE 8.17 Route to Trifluralin.

Other fluorine-containing aromatic substituents are most commonly located para to the other major functionality in the final product. Apart from SCF_3 and OCF_3, partially fluorinated side chains such as SCF_2Cl and, more recently, longer fluorinated chains such as $OCF_2CF_2CF_2H$, have value as substituents in aromatic products. As with most of the fluoroaromatic and benzotrifluoride products, the fluorine is commonly introduced early in the synthetic pathway, either as the first function in the molecule (many standard synthetic transformations, such as halogenations, nitrations, Friedel Crafts reactions, and reductions can be carried out on at least many of these aromatic fluorine compounds as discussed in earlier chapters) or via transformation (commonly involving halex or related chemistry) of one (or more) of several substituents on an aromatic ring.

REFERENCES

1. *Fluorine: The First Hundred Years (1886-1986),* R.E. Banks, D.W.A. Sharp and J.C. Tatlow, Elsevier, Lausanne, 1986.
2. *Agrochemicals — Executive Review,* 5th edition, Allan Woodburn Assoc. Ltd., Edinburgh, UK, 1994.
3. *The Pesticide Manual,* 8th edition, British Crop Protection Council, Bracknell, UK, 1987.
4. Chemistry International (IUPAC), 1994, 16, 210.
5. M. Silvester, *Performance Chemicals,* 1988, 18.
6. D.W. Ramsay, *J. Soc. Dyers Colourists,* 1981, 97, 102.
7. H. Burgin, Cibracon-C-Farbstoffe, paper read at the meeting of the Society of Northern Textile Engineers, Boras, Sweden, 1989.
8. S. Abeta and T. Omura, *Nikkakyo Geppo,* 1987, 8-15 (*Chem. Abstr.,* 1987, 107, 200367).
9. S. Abeta and K. Imada, *Rev. Prog. Coloration,* 1990, 20, 19.
10. J.P. Luttringer and A. Tzikas, *Textilveredlung,* 1990, 25, 311.
11. K.J. Herd, in *Organofluorine Chemistry: Principles and Commercial Applications,* eds. R.E. Banks et al. Plenum Press, New York, 1994, Chapter 13.
12. T. Inoi, in *Organofluorine Chemistry: Principles and Commercial Applications,* eds. R.E. Banks et al., Plenum Press, New York, 1994, Chapter 12.
13. *Organofluorine Chemistry: Principles and Commercial Applications,* eds. R.E. Banks et al., Plenum Press, New York, 1994.
14. P.E. Cassidy, T.M. Aminbhavi and J.M. Farley, *Rev. Macromol. Chem. Phys.,* 1989, C29(2&3), 365.
15. J.H. Clark and J.E. Denness, *Polymer,* 1994, 35, 5124.
16. J.W. Labadie and J.L. Hedrick, *Macromolecules,* 1990, 23, 5371.
17. W.R. Shiang and E.P. Wood, *J. Polym. Sci. Part A,* 1993, 21, 2081.
18. B. Minter, *Fluorine in Agriculture,* ed. R.E. Banks, Fluorine Technology Ltd., Cheshire, 1994, paper 3.
19. A.T. Woodburn, *Fluorine in Agriculture,* ed. R.E. Banks, Fluorine Technology Ltd., Cheshire, 1994, paper 2.
20. J.G. Dingwall, *Fluorine in Agriculture,* ed. R.E. Banks, Fluorine Technology Ltd., Cheshire, 1994, paper 20.
21. Y. Sanemitsu, *Fluorine in Agriculture,* ed. R.E. Banks, Fluorine Technology Ltd., Cheshire, 1994, paper 18.
22. R.E. Banks, *Fluorine in Agriculture,* ed. R.E. Banks, Fluorine Technology Ltd., Cheshire, 1994, paper 1.
23. G.T. Newbold, in *Organofluorine Chemicals and their Industrial Applications,* ed. R.E. Banks, Ellis Horwood, London, 1979.
24. S. Nathan, *Chem. Ind.(UK),* 1994, 936.

25. A. Marhold, *Fluorine in Agriculture,* ed. R.E. Banks, Fluorine Technology Ltd., Cheshire, 1994, paper 13.
26. W. Bernhagen, *Speciality Chemicals,* 1993, 256.
27. D. Cartwright, in *Organofluorine Chemistry: Principles and Commercial Applications,* eds. R.E. Banks et al., Plenum Press, New York, 1994, Chapter 11.
28. *Fluorine in Medicine,* eds. R.E. Banks and K.C. Lowe, UMIST, Manchester, 1994.
29. C. Heidelberger, N.K. Chaudhuri, P. Danneburg, D. Mooren, L. Griesbach, R. Duchinsky, R.J. Schnitzer, E. Pleven and J. Scheiner, *Nature,* 1957, 179, 1957.
30. A. Engel, *Organogluorine Chemistry: Principles and Commercial Applications,* eds. R.E. Banks et al., Plenum Press, New York, 1994, Chapter 13B.
31. C. Mercier and P. Youmans, *Fluorine in Agriculture,* ed. R.E. Banks, Fluorine Technology Ltd., Cheshire, 1994, paper 10.

Index